Bark *Catalpa* cutting in a sperm whale
  We soon cut him in
  And began to try out

Songs

The Whalemen Sang

by
Gale Huntington

Second Edition

Dover Publications, Inc., New York

This Dover edition, first published in 1970, is an unabridged republication, with minor corrections and a new Index of Song Titles, of the work originally published in 1964. It is reprinted by special arrangement with Barre Publishing Company, Inc., publisher of the original edition.

Standard Book Number: 486-22169-5
Library of Congress Catalog Card Number: 78-96995

Manufactured in the United States of America
Dover Publications, Inc.
180 Varick Street
New York, N.Y. 10014

ACKNOWLEDGEMENT

So MANY people have helped me with the making of this book that I could not name them all. But I am most grateful for every bit of help and information that I have received.

In particular I must mention all of the following whose assistance in one way or another made the book possible: Mrs. Ellen Chace, the librarian of the Nantucket Whaling Museum; Mrs. Eleanor Mayhew of the Dukes County Historical Society; Miss Norma Jean Lamb of the Buffalo and Erie County Public Library; Stuart C. Sherman, the librarian of the Providence Public Library; Laurence G. Hill, former librarian of the New Bedford Free Public Library; Philip F. Purrington, curator of The Whaling Museum of the Old Dartmouth Historical Society; M. V. Brewington of the Peabody Museum in Salem; Mrs. Laura Blikensderfer of the Kendall Whaling Museum; Edouard A. Stackpole, the director of the Old Mystic Seaport.

While I can not thank them all individually, the people in all the institutions where I have worked — both those mentioned above, and the following — were almost universally interested, helpful and kind: The Essex Institute of Salem; The Music Room of the Boston Public Library; The Harvard Music Library; The John Hay Library of Providence.

Also for encouragement, suggestions, or permission to use valuable material—or all three—I want to to thank A. L. Lloyd of Crooms Hill, London; Kenneth S. Goldstein of Hatboro, Pennsylvania; Iam Spink of the University of Sidney, New South Wales.; Edward D. Ives of the University of Maine; Gale Blosser of Milbrae, California; Mrs. Amos Smalley of Gay Head; Everett W. Whiting of West Tisbury; Joseph C. Allen of Vineyard Haven (all of Martha's Vineyard); and last, but certainly not least, my wife Mildred Tilton Huntington who helped me find songs, helped me copy songs, and helped me sing them.

The following permissions to use copyrighted materials are gratefully acknowledged: The Educational Company of Ireland, Dublin, for material from Joyce, *Ancient Irish Music* and *Old Irish Music;* the Harvard University Press for three melodies from Greenleaf and Mansfield, *Ballads and Sea Songs of Newfoundland;* The English Folk Song and Dance Society for melodies from *The Journal of the English Folk Song Society;* Chas. Taphouse & Son, Ltd., Oxford, for material from Kidson, *Traditional Tunes;* A. L. Lloyd for material from *The Penguin Book of English Folk Songs;* Ascherberg Hopwood & Crew, Ltd., London, for one melody from Kidson, *English Peasant Songs* and melodies from Kidson and Moffat, *A Garland of English Folk-Songs;* Corinth Books, Inc., New York, for melodies from Lochlainn, *Irish Street Ballads;* The American Folklore Society for material from *The Journal of American Folklore* and *The Bulletin of the Folk Song Society of the Northeast;* Alan Lomax for material from Lomax and Lomax, *American Ballads and Folk Songs;* The W. W. Norton Co., New York, for melodies from Colcord, *Songs of American Sailormen;* Barre Publishers, Barre, Massachusetts, for material from Harlow, *Chanteying Aboard American Ships.*

CONTENTS

ACKNOWLEDGEMENT   *page* v

CONTENTS   vii

LIST OF ILLUSTRATIONS   xiii

INTRODUCTION   xv

SONGS OF WHALES AND WHALING   1
  The Coast of Peru   2
  A Prayer at the Start of a Voyage   7
  A Fitting Out   7
  The Whalefish Song   9
  The Greenland Whale   11
  The Cruise of the Dove   13
  The Whaleman's Lament   15
  The Whalers' Song   17
  A Whaling Scene   21
  Our Old Friend Coffin   22
  The Wounded Whale   23
  Rolling Down to Old Mohee   27
  Diego's Bold Shore   30
  The Old Hulk   32
  The Bark Gay Head   34
  The Bark Ocean Rover   37
  Desolation   38
  The Wings of a Goney   40
  Blow Ye Winds   42

OF THE SEA AND SHIPS   47
  The Ship Euphrasia   47
  A Wet Sheet and a Flowing Sea   49

Sling the Flowing Bowl 51
Loose Every Sail to the Breeze 52
Captain James 54
The Topsail Shivers in the Wind 59
The Sequel to Will Watch 62
The Sea 63
Saturday Night at Sea 65
I Was Once a Sailor 66
Hearts of Gold 68
The Tempest 70
The Can of Grog 73
The Pirate of the Isles 74
The Demon of the Sea 78
The Rover of the Sea 80
Most Beautiful 81
The Sea Ran High 81
The Ocean Queen 82
The Storm Was Loud 83
Neptune 83
The Dauntless Sailor 85
The Sovereign of the Sea 85
A Life on the Ocean Wave 87

OF SAILORS AND MAIDENS FAIR 90
Covent Garden 90
Cupid's Garden 92
William Taylor 94
The Tarry Trousers 96
The Captain Calls All Hands 99
A Young Virgin 100
The Nobleman's Daughter 103
John Riley 105
The British Man-of-War 108
Pretty Sally 111
The Banks of Glenco 113
The Beggarman 116
Bright Phoebe 119
The Dark-Eyed Sailor 120
The Mantle So Green 122

Our Ship She is Lying in Harbour 124
The Silvery Tide 125
The Undutiful Daughter 127
The Ship Carpenter 129
The Pride of Kildare 131
The Lily of the West 133
The Maid on the Shore 136
Lovely Caroline 137
The Turkish Lady 141

OF YANKEE MANUFACTURE 144
The Times 144
The Sons of Liberty 146
The Lass of Mowee 148
The Heathen Dear 151
The Moon is Brightly Beaming Love 152
Shearing Day 153
Sarah Mariah Cornell 156
John Brown 158
The Banks of the Schuylkill 160
The Banks of Champlain 161
A New Liberty Song 163
A Song on the Nantucket Ladies 165
Springfield Mountain 167
Blessed Land of Love and Liberty 170
Sons of Worth 171
John Bull's Epistle 172
The California Song 174
The Captain 176
The Confession 179
Virtuous America 180
The Indian Hunter 180
Willie Gray 182

ALL FROM THE BRITISH ISLES 184
The Shepherd's Daughter 185
A New Song 187
The Croppy Boy 188
Queen of the May 190

The Sheffield 'Prentice Boy | 192
Song on Courtship | 194
When First into this Country | 195
O Logie O Buchan | 197
Aran's Lonely Home | 198
Fair Betsy | 201
I Had a Handsome Fortune | 203
Bonaparte on St. Helena | 205
The Bonny Bunch of Roses-O | 207
Bonaparte | 209
The Green Linnet | 211
One Night Sad and Languid | 215
The Farmer's Boy | 216
The County of Tyrone | 218
In Days When We Went Gipsying | 220
Rinordine | 222
Behind the Green Bush | 223
Reily's Jailed | 224
The First Time I Saw My Love | 225
The Shepherd's Lament | 227
Women Love Kissing As Well As the Men | 228
Fanny Blair | 229

PARLOR SONGS THAT WENT TO SEA | 232
Silvery Moon | 233
Willie's on the Dark Blue Sea | 234
The Banks of Banna | 236
The Maid of Erin | 237
Adieu My Native Land | 238
Angels Whisper | 239
The Bride's Farewell | 241
The Dying Soldier | 243
Genette and Genoe | 245
Mary's Dream | 246
The Ocean | 248
Thou Hast Learned to Love Another | 249
We Met 'Twas in a Crowd | 251
Jamie's on the Stormy Sea | 252
Adieu to Erin | 255

Blow High Blow Low   256
The Rose of Allendale   257
The Beacon Light   260

FRAGMENTS   262
A New Sea Song   262
Down Wapping   263
The Bible Story   264
Farewell My Dear Nancy   266
The Turkey Factor in Foreign Parts   268
The Sailor Boy's Song   271
A Sailor's Trade is a Roving Life   272
Hunter's Lane   273
Fare You Well   274
Moll Brooks   274
Nelson   275
The Bonnet of Blue   275
When I Remember   277
Come Let Us Be Jolly   277
Ye Parliaments of England   278
Old Horse   279

LAST BUT NOT LEAST   282
An Ancient Riddle   282
Wait for the Wagon   285
Prayer   286
The Pilot   288
The Lord Our God   289
Row On   290
The Recruiting Sargeant   291
The Post Below   292
A Love Song in the Year 1769   293
Elegy on the Death of a Mad Dog   295
A Charming Fellow   296
The Wreath   297
I Can Not Call Her Mother   298
The Village Born Beauty   299
As I Grow Old   300
Norah Darling   302

The Sailor's Farewell   304
Terrible Polly   306
The Wide World of Waters   308
Song of Solomon's Temple   309
Poll and Sal   312
The Keyhole in the Door   315
The Sandshark   317
Billy O'Rourke   318
Now We Steer Our Course for Home   321

INDEX OF SOURCES   323

BIBLIOGRAPHY   326

INDEX OF SONG TITLES   329

LIST OF ILLUSTRATIONS

ALL ILLUSTRATIONS ARE REPRODUCED THROUGH THE COURTESY OF
THE WHALING MUSEUM, NEW BEDFORD, MASSACHUSETTS

Bark *Catalpa* cutting in a sperm whale
  We soon cut him in
  And began to try out   *frontispiece*

Ship *Atkins Adams* from a journal page
  The Larboard boat got one
  And the bow boat another   18

Scrimshaw on sperm whale's tooth
  And if you get too near his flukes
  He'll kick you to the devil   29

Whale stamps from journal page (each stamp in the ship's log
meant a dead whale)
  Bowheads and blackskins   44

Page from a journal
  We have the noble waist boat
  Whose crew are very good   61

Bark *Stella* from a journal page
  It's all about a whaling bark
  That left New Bedford city   76

From a journal page
  Sometimes you ship a sea
  Sometimes you see a ship   86

Scrimshaw on a sperm whale's tooth
  You scarce could find so fair a dame
  To search this wide world over   102

Scrimshaw on matched sperm whale's teeth
  The first time I saw my love, happy was I   118

From a journal page
  And when I'm gone love think of this
  When will we meet again?   140

Scrimshaw on a sperm whale's tooth
  I espied a fair damsel
  She appeared like a queen   155

Scrimshaw on a sperm whale's tooth
  We'll drink and sing
  While foaming billows roll   168

Scrimshaw on a sperm whale's tooth
  Come all you jovial sailors
  That love your native home   178

Scrimshaw on a sperm whale's tooth
  He has been a gallant voyage
  And has lately come on shore   214

Scrimshaw on a sperm whale's tooth
  Fanny Blair is a girl of eleven years old   226

Carving on whale or walrus ivory
  There was something in your glances
  Put a summer in our fancies   242

Scrimshaw on a sperm whale's tooth
  We worked for our lives
  While each tar done his best   254

From a journal page
  Lay on Captain Bunker
  I'm hell for to dart   259

From a journal page—Bark *Herald* trying out
  To catch the whales
  And cut and boil   269

Scrimshaw on a sperm whale's tooth
  We'll tow him alongside
  And rob him of his hide   280

From a journal page—three whalers gamming
  And if you should gam her
  Just bear it in mind   294

Scrimshaw on a sperm whale's tooth
  Now our boats being lowered there arose a contest
  Among the boats crews t'see which should do best   310

Enlarged section of scrimshaw art on whale ivory
  The men were sprawling in the sea
  And swimming for their lives sir   320

INTRODUCTION

For MORE than two hundred years New England and the sea were
synonymous. The wealth of New England came from the sea, or
from trade on the sea, and although old England would never admit it,
New England ships and New England seamen became the best in the
world.

Gradually, as the years passed, for some reason or another, particular
towns and ports and areas became identified with different aspects of
seafaring — Salem and Boston with merchant voyages to all the far
ports of the earth; Gloucester and Cape Ann with fishing; Newburyport
with sugar and rum which figured so importantly in the triangular trade;
Rhode Island, sad be it, with the slave trade; Maine with coasting; and
southeastern Massachusetts with whaling.

That is very much a generalization, for all the New England coast
had its fishermen, and every port, no matter how small, its coasting vessels. Boston as well as the Rhode Island towns gained wealth from the
slave trade. And whaling vessels sailed from many towns besides those of
southeastern Massachusetts. But still the generalization does hold good.
Gradually the little seaport towns of the area just south of Cape
Cod became the center of the New England whaling industry, and finally
the center of the whaling industry of the world.

Nantucket was first and then New Bedford far outpaced her. Martha's Vineyard, while it never had a whaling fleet to compare in size
with those of its two great rivals, always furnished far far more than its
share of men and masters, mates and boatsteerers for the ships of both
of them.

Why did this specialization take place? Why did this small area
turn to whaling in particular and become great at it, and wealthy? There
must be many reasons, for certainly no one of them alone furnishes the
answer. In the early days whales were very plentiful in these waters, but
they were plentiful on other parts of the New England coast, too. The

INTRODUCTION

Indians of this area had long been great whalemen going far to sea in their big dugout canoes. But Indians of other parts of the coast also went whaling.

But no matter what the reason or reasons, it was in the little towns on Nantucket and Martha's Vineyard, and about the shores of Buzzards Bay that whaling became the established industry, and it was there that boys growing up knew from the time that they were able to pull an oar, and that was pretty young, too, that they were going to be whalemen. At thirteen and fourteen and fifteen those boys went to sea, and they are the ones mostly, who kept the journals that furnish the ballets for this book.

In the very earliest days whaling was from the beach, Indian fashion. The only difference being that the Indians ate the whale and the Yankees tried him out for his oil. Then as whales became scarcer near the shore the whalemen had to cruise to sea for them. At first these whaling vessels were very small and they carried the whale's blubber stowed raw in the hold, and the oil was still tried out on the beach. But if only a few days of adverse weather prevented a quick return, the blubber in the hold became a putrid mess. So the next step was to build a try works on deck. That, essentially, was nothing but a brick platform where a fire could be made safely, and the oil from the blubber cooked out in big iron pots. Then the oil would be stowed below decks in wooden casks where it would keep sweet indefinitely. That was the real beginning of pelagic whaling in New England.

And still the voyages were not too long. It was only toward the end of the eighteenth century when vessels began to go around Cape Horn and into the Pacific after whales that voyages began to last a year and more. Finally voyages of two and three and even four years became common. And that was a long time to be away from home. It was then that keeping a journal of the voyage could, and did, become a solace and comfort.

In the beginning whaling was incidental to a man's regular occupation which might be farming or fishing or coasting. But eventually it developed into a profession and a very hard and rigorous one.

The captains and mates were the masters of their trade, the true professionals. The petty officers, the boatsteerers, were the journeymen, and the boys who pulled at the oars were the apprentices. But as whaling grew and became a great industry with hundreds of vessels fishing, and all of them with large crews, there just were not enough boys to go around, and so, " 'Twas advertised in Boston, New York, and Albany" for men to go whaling. And still there were not enough, so Portuguese sailors were signed aboard from the Azores and Madeiras and Cape Verde Islands.

Thus it came about that on the whaleships there were two classes, the two professionals, mainly from southern New England, and the others. Some of the others did become professionals, but except for the Portuguese, one voyage was usually more than enough for most of them. Mainly it was the professionals who kept the journals and log books.

Usually it was the charge of the first mate to keep the official record of the voyage — the log book. His entries were almost universally brief, factual and dull. Often other members of the crew would keep their own records of the voyages, too, and those are the journals. As most of the journals are quite as dry and dull as the log books it is often very hard to tell them apart.

But a few of the journals are something else again, and written with feeling and perception. Some are decorated with pen and ink line drawings, and a few with quite lovely water color sketches. It is in the journals, but only in about one in seventy of them, that one finds most of the songs.

Now why were these songs recorded in the journals? The answer to that is easy. Music — song — was one of the very few real pleasures that the whalemen had. Often the food was very bad, sometimes the officers were cruel and brutal, and always the work was dangerous and hard. Add to that the fact home and everything that home stood for might be thousands of miles, and years away, too, and you will see why most of the whalemen sang, and why some few of them recorded the songs that they sang and loved.

What were the songs that they sang? Why, everything that was sung on shore, of course, plus the special songs that had to do with their trade. They sang of storm and shipwreck, of love and laughter and tragedy and death. Some of their songs were the popular songs of the period of the voyage. Others, many of them, were the old, old songs and ballads from the British Isles. There were gospel songs and music hall songs, parlor songs and bawdy songs. The only songs that were not recorded in the journals are the chanties. And that is because the whalemen, like all seamen, did not think of the chanties as songs at all. They were a part of the routine of working ship and everyone knew them. Besides that, it was always considered bad luck to sing them apart from the work.

So, the songs in this book are all from ballets in journals, log books, and two manuscript collections of songs. Almost all were recorded by whalemen. A few, particularly some of the earliest, were recorded on merchant voyages, but not, I think, enough to change the title of the book, the more so as in the early days of whaling most whalemen also made merchant voyages.

Again, the melodies in this book have no scholarly interest or value. However, each and every one of them is a melody that was actually used with a version of the song that it accompanies. And that is all I can say for them, for when William Histed sang "Down Wapping," or when Ira Poland sang "The Queen of the May," they may have used entirely different tunes than the ones I have included here. But the tunes that I have included for "Dawn Wapping" and "The Queen of the May" were traditional for them. And so it is with all the other songs in the book.

One could wish that the whalemen had included music for their songs, but they didn't, and so I have had to supply it. For a folk song without a melody is a poor, broken thing. The melodies that I have used come from every possible source; from old song books, books of fiddle tunes, sheet music, and scholarly works. Occasionally I have had to change and trim a melody to make it fit the particular ballet. But I have done as little of that as possible.

To keep the book uniform, the songs have been left entirely without punctuation, for that is how most of them are found in the journals. Also, most of the spelling, which was often very bad indeed, has been corrected, leaving only such few misspellings as seem to add to the meaning or interest of the songs. Most uncalled for capitalization has been changed to lower case, and obvious mistakes have been corrected.

And so I hope that this book may prove interesting and valuable not only to folk singers like myself, but also to serious folk song scholars, and to students of Americana. For whaling is one of the great maritime traditions of our people and these are the songs that the whalemen sang.

SONGS

THE WHALEMEN SANG

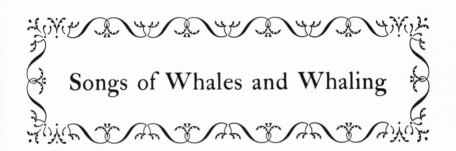

Songs of Whales and Whaling

A LL OF the craft of whaling is in these songs — the excitement, the danger, the hope of financial gain, and the heartbreak.

Whaling was many things. For more than two hundred years it was a great industry, for part of that time it was the greatest single industry that New England possessed. It was also adventure of the highest order. Pride of craft was high in the whalemen. They thought they were the finest seamen on earth and perhaps they were. They thought that they were the finest boatmen, too, and they were indeed just that without any qualification whatsoever.

Although whaleships, at one time or another sailed from many American ports, the center of the industry was always that small area that included the Islands of Martha's Vineyard and Nantucket, and the little towns about the shores of Buzzards and Narragansett Bays. It was almost exclusively from there that the best of the officers and boatsteerers were drawn.

In the course of their long voyages the whalemen contributed to the sciences of navigation and oceanography. Many of the islands of the Pacific were discovered and first recorded on charts by them. They recorded ocean currents and meteorological phenomena. And they brought back knowledge of the far places of the earth, and much of this is in the songs that they sang.

So here are some of the songs of whaling. I wish that there w
more, but many, I am afraid, have been lost forever.

1

THE COAST OF PERU

Come all you young fellows
That's bound after sperm
Come all you young fellows
That's rounded the Horn
Our captain has told us
And we hope it will come true
That there's plenty of sperm whales
On the coast of Peru

We have weathered the Horn
And are now on Peru
We are all of one mind
And endeavor to do
Our boats are all rigged
And our masthead all manned
Our riggin' rove light
And our signals all planned

first whale we saw
s late in the day
aptain come up
hese words he did say
to your hammocks

2

And quiet there be
We will see him in the morning
Close under our lee

Next morning at daybreak
About five o'clock
The man at the masthead
Cried yonder she spouts
Where away does she lay
And the answer from aloft
Two points on our lee bow
And about three miles off

Then it's call up all hands
And it's be of good cheer
Put your tubs in your boats boys
Have your bow lines all clear
Sway up your boats now
Jump in you boat's crew
Lower away now lower away
My brave fellows do

Our waist boat got down
And of course got the start
Lay on Captain Bunker
I'm hell for to dart
Now bend to your oars boys
And make the boat fly
But one thing we dread of
Keep clear of his eye

Now the chief mate he struck him
And the whale he went down
And the captain pulled up
And he tried to bend on
But the whale began to vomit
And blood for to spout

And in less than ten minutes
We had him fin out

We towed him alongside
And with many a shout
We soon cut him in
And began to try out
Now our whale she is tried
And likewise stowed down
She is better to us boys
Than five hundred pounds

Here's a health to all whalemen
Boys drink it down do
Likewise to the Bengal
And our whaleship's crew
All you that want money
I would have you to go
On the coast of Peru
Where the sperm whales do blow

Bengal 1832

This song probably dates back to the last quarter of the eighteenth century when whaling in the Pacific was still new. The *Bengal* is named in the last stanza, but that has no special meaning, for often the vessel on which a song was sung would be named. The captain told the men to get into their hammocks, and that too is an indication that the song is old, for later whaling vessels were all equipped with bunks. And the captain's name was Bunker. Bunker is one of the old family names of Nantucket, and has always been identified with whaling.

This is a good song. The hope and high expectation are here, the excitement of the chase and the kill, as well as many of the technical details. Very briefly I shall try to explain some of the latter.

"Our boats were all rigged" means that the whaleboats were swung on the davits ready for lowering, and that the gear was stowed. The amount of gear carried in a whaleboat is astounding. First of all there were the oars, three for the starboard and two for the larboard sides as well as the boatheader's great steering sweep. And there were spare oars in case of breakage or loss overboard. Besides the oars there were paddles, perhaps a direct hangover from the days of Indian whaling. And if the sea was very quiet the boat would approach the quarry with paddles rather than oars, for the paddles made less noise.

Then there was the boat sail wrapped around its mast. It was usually a spritsail, but in the earlier days of whaling a lug sail was used. And after the centerboard was invented — perhaps about the beginning of the nineteenth century — almost all whaleboats had a centerboard. And besides all that gear there was the assortment of harpoons and lances and spades for striking and killing the whale. There were buckets for bailing and for wetting down the line as it turned around the loggerhead, and the rudder and tiller which were used rather than the steering sweep if the boat had to travel any distance. There would probably be a drogue or sea anchor. It's hard to see how the men could make out, so cluttered was the boat with gear, but they did.

"Our signals all planned." Every whaleship worked out its own set of signals which it was hoped would be unintelligible to other whalers on the grounds. These were signals from the ship to boats and the other way around. They would include, whale sighted, whale struck, dead whale, whale working to windward — or whatever other direction — and many more. Some of the logbooks and journals contain the list of signals that were used for that particular voyage.

"The man at the masthead cried yonder she spouts." A watch from the masthead was kept for whales during all the daylight hours, all men taking their turn at this duty. And a whale could be recognized by its spout as a sperm whale, a right whale, bowhead, blackfish, and so on. When the whale was sighted the man who sighted it would cry out, "Blows, there she blows." The "there" usually came out "thar" and so we have the familiar, "Thar she blows."

"Where does she lay?" When the whale was sighted the captain or the officer on deck would call, "Where away?" or "Where does she lay?" And the answer would come back from aloft, "Two points on our lee bow," or wherever. Then the captain would usually go aloft himself to verify the sighting.

Then the boats were lowered and the chase was on. Each boat was commanded by an officer — the captain or one of his mates. The number of boats would depend on the size of the vessel. A small schooner might carry only two boats and would have but one mate. If there were three boats there would be two mates and so on. In the heyday of whaling the usual complement was five boats and four mates. There was always a great rivalry among the boats to be first on the whale.

"Now the chief mate he struck him." That means that the boatsteerer in the chief mate's boat had struck the whale — had driven in the harpoon. The boatsteerer rowed forward oar usually called harpoon oar, on the starboard side, and he would ship his oar to be ready to strike when the officer told him they were on — approaching — the whale. As soon as the harpoon was in the whale the boatsteerer would go aft and change places with the captain or mate, and steer the boat — hence boatsteerer. Then it was the officer's job to go forward and kill the whale, or try to, as soon as the opportunity offered.

It seems like a complicated and unnecessary process, but that is the way it was. Killing the whale was not only the most dangerous job but also the most important, and thus it was the captain's or mate's duty to perform it.

"The captain pulled up and tried to bend on." This means that the whale had sounded — gone down — and taken most of the line from the tubs in the chief mate's boat with him. So the captain was trying to make his line fast — bend on — to the chief mate's line before it was gone and the whale lost. Sometimes the lines from all the boats would be bent on before the whale had had enough of it and had to return to the surface to breathe. Then the boats would converge on him to try for the kill.

"We had him fin out." The whale was dead.

"We towed him alongside." Alongside the ship, where he would be made fast fore and aft with heavy cables. Then the cutting stage would be lowered over the whale, and the process of removing his blubber — cutting in — would begin. That was the boatsteerers' and mates' job, and a slippery, dangerous, toilsome one it was. Razor sharp cutting spades were used in this work, and it was said that very few whalemen reached old age with all of their toes.

"Now our whale she is tried out." The oil was boiled — tried or fried — from the blubber in great iron pots set in the tryworks, a structure and platform of brick that was built on deck. The fire would be started with wood kept for that purpose, but after it was going it was kept alive with the pieces of blubber from which the oil had already been extracted. When the ship was full the try works would be knocked down and thrown overboard and a new one would be built when the ship was fitting out for her next voyage.

See Colcord, pp. 194-195; Harlow, pp. 222-223; Williams, Alfred M., pp. 33-35.

A PRAYER AT THE START OF A VOYAGE

God keep these cheery mariners
And temper all the gales
That sweep against the rocky coast
To their storm shattered sails
And men on shore will bless the ship
That could so guided be
Safe in the hollow of His hand
To brave the mighty sea

Rebecca Sims 1854

Many of the journals begin with a prayer. This little eight line prayer from the *Rebecca Sims* journal suggests quite a lot about the whalemen. There were godless men aboard the whaleships to be sure, but there were God-fearing men, too.

A FITTING OUT

A chest that is neither too large nor too small
Is the first thing to which your attention I'll call
The things to put in it are next to be named
And if I omit some I'm not to be blamed

Stow first in the bottom a blanket or quilt
To be used on the voyage whenever you wilt
Thick trousers and shirts woolen stockings and shoes
Next your papers and books to tell you the news

SONGS THE WHALEMEN SANG

Good substantial tarpaulins to cover your head
Just to say keep it furled N. C. nuff said
Carry paper and ink pens wafers and wax
A shoemaker's last awls and some small tacks

Some cotton and thread silk needles and palm
And a paper of pins as long as your arm
Two vests and a thimble a large lot of matches
A lot of old clothes that will answer for patches

A Bible and hymn book of course you must carry
If at the end of the voyage you expect for to marry
Don't forget to take esseners pipes and cigars
Of the sweetest of butter a couple of jars

A razor you will want a pencil and slate
A comb and a hairbrush you will need for your pate
A brush and some shaving soap and plenty of squills
And a box of those excellent Richardson's pills

A podeldoe and pain killer surely you will need
And something to stop the red stream should you bleed
Some things I've omitted but never mind that
Eat salt junk and hard bread and laugh and grow fat

Ocean Rover 1859

Fitting out for a whaling voyage that might last three or more years was a serious business, and the things that went into the chest had to be carefully selected, for a chest would hold only so much. What esseners were, or the podeldoe, I have no idea.

A good question here is, would verses like these actually be sung? And the answer is yes. I can remember my wife's grandfather, Welcome Tilton, making up verses about some event that had just taken place — any sort of event. He always seemed to make up the verses to whatever tune was going around in his head at the moment. And some tune was always going around there.

THE WHALEFISH SONG

In eighteen twenty four we sailed
On June the twentieth day
When we jolly tars our anchors raised
And away for the whalefish steer brave boys
And away for the whalefish steer

Bold Stevens was our captain's name
And our ship the lion bold
We had fourteen men and they were brave
To face the wind and cold brave boys
To face the wind and cold

It was when we came to that country cold
Our goodly ship to moor
It was then we wished ourselves back again
With our pretty girls on shore brave boys
With our pretty girls on shore

Our boatswain he goes up aloft
With a spyglass in his hand
Overhaul overhaul on our davit tackle fall
And launch our boat to the sea brave boys
And launch our boat to the sea

Our captain on the quarter deck stands
And a clever little man was he
It's a whale oh a whale and a whalefish he cries
And he blows at every span brave boys
And he blows at every span

Cheer up cheer up my lively lads
Let not your courage fail
For Providence will have her way
Let a man do all he can brave boys
Let a man do all he can

Here is success to the Lion bold
And to all her jovial crew
Here's a health to our pretty girls at home
And your married men's wives on shore brave boys
And your married men's wives on shore

Euphrasia 1849

There are almost as many names for this song as there are versions of it. Sometimes it is called "Greenland" and sometimes "The Greenland Whale" and the "Greenland Whale Fishery." It is also called just "The Whale," and sometimes "The Whale Catchers." And there are more, many more.

Originally this song was English and dates back at least to the beginning of the eighteenth century, and perhaps even much earlier than that, for there were English and Dutch whalers on the Greenland ground early in the sixteenth century. And American vessels were "on Greenland" in early colonial times. This version, as well as the one that follows, is purely English for American whaling vessels never carried a boatswain.

The line "We had fourteen men and they were brave," must refer to foremast hands only, and even so, the *Lion* was a very small vessel. The *Lion*, by the way, is the name of the vessel in almost all versions. This song has been collected over and over again, and still it is about as up-standardized as any folk song can well be. See *The Book of Navy Songs*, p. 76, and p. 120; Vaughan Williams and Lloyd, pp. 50-51, and

p. 100; Colcord, pp. 151-152; Whall, pp. 71-73; Baring Gould, pp. 56-57; and JFSS, vol. 1, pp. 101-102, vol. 2, p. 243, and vol. 7, p. 228.

THE GREENLAND WHALE

It was seventeen hundred and eight-four
On March the seventeenth day
We weighed our anchor to our bow
And for Greenland bore away brave boys
And for Greenland bore away

Bold Stevens was our captain's name
Our ship called the *Lion* so bold
And our poor souls our anchor away
To face the storms and cold brave boys
To face the storms and cold

Oh when we arrived in that cold country
Our goodly ship to moor
We wished ourselves safe back again
With those pretty girls on shore brave boys
With those pretty girls on shore

Our boatswain to the main top stand
With a spyglass in his hand

11

SONGS THE WHALEMEN SANG

A whale a whale my lads he cries
And she spouts at every span brave boys
And she spouts at every span

The captain walked the quarter-deck
And a jolly little fellow was he
Overhaul overhaul your davit tackle falls
And we'll launch our boats all three brave boys
And we'll launch our boats all three

There was harpineery and picaneery
And boat steerery also
And twelve jolly tars to tug at the oars
And a-whaling we all go brave boys
And a-whaling we all go

We struck that whale and down she went
By the flourish of her tail
By chance we lost a man overboard
And we did not get that whale brave boys
And we did not get that whale

When this news to our captain came
It grieved his heart full sore
And for the loss of a 'prentice boy
It was half mast colors all brave boys
It was half mast colors all

It's now cold months is a-coming on
No longer can we stay here
For the winds do blow and the whales do go
And the daylight seldom does appear brave boys
And the daylight seldom does appear

Bengal (2) 1833

This song is from the *Bengal* journal (2) which was kept by William Silver of Salem. It is another of the many versions of "The Whalefish Song." Here again the British touch is seen: in American vessels it was almost always the captain who went aloft when a whale was sighted leaving the mate on the quarter-deck. Also the captain almost always went with the boats, one of which was his.

Mooring the vessel and using her as a factory ship, the whales being towed to her by the boats, was practiced in other parts of the world besides Greenland, notably in the Marquesas and off the coast of Lower California. And American vessels did not carry apprentices as such. All the boys in the crew were learning the trade. The "Twelve jolly tars for to tug at the oars" would mean — if the *Lion* were an American vessel — that she had three boats.

Of course when winter came on the Greenland fishery was closed down until the beginning of another season.

THE CRUISE OF THE DOVE

Ye men of renown who are a-swearing
For some noble deeds your lordships have done
Come listen to me and hear things full as daring
Which we brother spouters think nothing but fun

It was a fine ship with prime captain and crew
Surpassed by none and equaled by few
With courage undaunted by oars and by sail
So nimble we chased the spermacety whale

The name of our ship I suppose you'd like to know
The name of our captain and owners also

SONGS THE WHALEMEN SANG

She was called the Dove you will see in my song
And nothing I tell you I swear it is wrong

Our captain's name is Butler a man fine and bold
Our owners named Hazzard and Worth I hear told
On the Coast of Peru we were destined to cruise
But if we'd stopped there it would not have been much use

Then away to the westward in hopes for to find
Some work for all hands being that way inclined
Then away to the northward to Japan likewise
Where the whales both the irons and lances defies

There she breathes there she blows was the cry heard one day
The captain looked up and hailed out where away
Right ahead and abeam on each hand them we spy
Like logs around us so sweetly they lie.

The captain went up and he soon gave the word
Being his orders by all should be heard
So back the main yard and stop the ship's way
Do swing out your boats boys and lower away

Now our boats being lowered there arose a contest
Among the boats crews to see which should do best
Spring on says the headsman don't let them pass by
When up starts a whale and lay on is the cry

Stand up was the next word that I heard him say
Into her she's got it lay on the other way
I have got a good iron just over her fin
So work sharp my boys and pull on her again

Now we worked for our lives while each tar done his best
We brought the school to and had work for the rest
And while that our whales were bleeding and dying
The shipkeepers so anxious were to windward ever plying

Now our whales are turned up and we prepared for our toil
We will soon get on board with the blubber to boil
When it's boiled out and stowed down in the hold
We'll drink greasy luck to the whalers so bold

Our ship she is full and home we are bound
We fill up our glasses and drink all around
We fill up our glasses and so merry we will be
And drink a good health to the liberty tree

Now in New York harbor our good ship lies moored
With a hold full of oil and all hands well on board
Being paid by our owners we leave captain and mate
We're bound for the park boys to blow us out straight

Minerva 1845

This is a very old song, perhaps even older than "The Coast of Peru" for the *Dove* was one of the very first Yankee whaleships to fish in the Pacific in the last quarter of the eighteenth century. The *Dove* was sloop rigged and she hailed from Nantucket. I cannot find any record of where she unloaded oil in New York, but she may well have done so as the song states. She did unload at least one voyage at Newport, Rhode Island, and the owners for that voyage were Hazzard and Worth.

The shipkeepers in the tenth stanza were those who stayed on the vessel and worked her while the boats were out after whales. They were: the cook, the steward, the cabin boy, as well as the carpenter, cooper, and sail-maker. Sometimes they had their hands full working the ship to keep the boats and whales in sight.

"Greasy luck" in the eleventh stanza was the standard toast of whalemen. I have heard the term used over and over again by old whalemen on the Vineyard. The melody I have used here is the "Liberty Tree." I feel pretty sure that that is the tune that was used.

THE WHALEMAN'S LAMENT

15

'Twas on the briny ocean
On a whaleship I did go
Oft times I thought of distant friends
Oft times I thought of home
Remembering of my youthful days
It grieved my heart full sore
And fain I would return again
To my own native shore

Though dreary discontented
I then resolved to (go)
My fortune on the seas (to try)
To ease me of my woe
I shipped me on a whaleship
To sail without delay
To the Pacific Ocean
There for a while to stay

Through dreary storms and tempest
And through some heavy gales
Around Cape Horn we sped our way
To look out for sperm whales
They will rob you they will use you
Worse than any slaves
Before you go a-whaling boys
You had best be in your graves

They'll flog you for the least offense
And that is frequent too

And the best that you will get from them
Is plenty more work to do
So do it now or damn your eyes
I will flog you till you're (blue)
My boys I wouldn't say it all
But it is all too true

But if I ever return again
A solemn vow I'll take
That I'll never go a-whaling
My liberty to stake
I will stay at home
And I will roam no more
For the pleasures are but few my boys
Far from our native shore

Catalpa 1856

Whether or not the crew was as badly treated as this song indicates depended entirely on the afterguard and usually on the captain. Some whalers were happy ships and others just plain were not. Then, too, whalemen were divided into two classes: the first comprised those who were born to go whaling and who knew they were going to be whalemen from the time they were old enough to pull an oar; the other class came from the cities and back country and had no tradition of whaling in their blood. It was men of that latter class who so often had a bad, bad time on a voyage.

THE WHALERS' SONG

Ship *Atkins Adams* from a journal page
The Larboard boat got one
And the bow boat another

There she lies there she lies
Like an isle on ocean's breast
Where away west south west
Where the billows meet the skies
Port the helm trim the sail
Let us share this mighty whale
There she blows there she blows
Man the boats for nothing stay
Such a prize we must not lose
Lay to your oars away away
Give way careful steer
Launch the harpoon laugh at fear
Plunge it deep the barbed spear
Strike the lance in swift career

Give her line give her line
Down she goes through the foaming brine
Sponge the side where the flying coil
Marks the monster's speed and toil
But though she dives to the deepest ground
Where the lead line fails to sound
Where the coral gardens hide
'Tis all in vain 'tis all in vain
She hath that within her side
That will bring her up again

Spout spout spout
The waves are purling all about
Every billow on its head
Strangely wears a crest of red
See her lash the foaming main
In her flurry and her pain
Take good heed my hearts of oak
Lest her flukes as she lies
Swiftly hurl you to the skies
But lo her giant strength is broke
Now she turns a mass of lead
The mighty mountain whale is dead

Row row row
In our vessel she must go
Changed into a liquid stream
O'er the broad Pacific's swell
Round Cape Horn where the tempests dwell
Many a night and many a day
Home with us she must away
Till we joyful hail once more
Old Nantucket's treeless shore
And the fair whom we please
Sits by the fireside at her ease
Let her remember if she will
The hardy tar who on seas afar
Risked his life her lamp to fill

Lexington 1853

This song is evidently pretty badly garbled, and yet there is so much power and beauty in it that I felt it should be included here. The melody I have used is "Hearts of Oak" or "The Liberty Tree," and it seems to fit pretty well. After that first stanza, though, anyone who wants to sing it will be on his own and will have to do plenty of arranging. But that, of course is a folk singer's privilege — not a folk song collector's.

"Let us share this mighty whale" refers to the fact that whalemen were never paid a wage. They received a share or "lay." The lay

depended on the rank or position held. The captain's lay was the largest
and the cabin boy's the smallest.

A WHALING SCENE

A whaling scene I'll now relate
On ocean's bosom wide sir
In boats away far from their ship
In chase of shoals of whales sir

The headsman shouts with arm uplift
With harpoon to let drive sir
And in the body of the fish
He sets it firmly in sir

Now when the wound the whale doth feel
He scampers off quite quick sir
Perhaps he gives the boat a whack
In pieces soon it is sir

The men were sprawling in the sea
And swimming for their lives sir
Until another boat shall come
And give them help quite soon sir

The pursuit then is begun again
With . . . . . .   ?   . . . sir
And soon they haul up to his back
And pierce him well and firm sir

The blood now soon begins to spout
Look out for fins and flukes sir
And after blustering for some time
His flukes he soon will peake sir

Now this advice 'tis well to take
To place your iron firm sir

21

And look out sharp and keep quite clear
From fins and flukes and jaws sir

Maria (2) 1846

Both this and "Our Old Friend Coffin" which follows, are credited to William Carroll, who must have been an American living on St. Helena. Perhaps he was a factor, or a ship chandler or in the consular service. At any rate, in spite of the use of the lazy man's device of "sir" in place of a rhyme, he does tell us something about whaling in both songs.

I am ashamed to say that I could not decipher the second line of the fifth stanza. Those who must have a song complete may supply their own words. Or, "with all the boats in line sir" would do.

OUR OLD FRIEND COFFIN

Our old friend Coffin is now here
From off a voyage good sir
'Tis many years since we did see
This worthy friend of ours sir

The cargo he has now on board
We hope will sell quite high sir
A noble price he ought to get
For all his toil and work sir

To catch the whales and cut and boil
To do so is no joke sir
If oil was California gold
It would not be so much sir

These whalemen brave their work quite harsh
Do nobly earn their dues sir
If oil would bring a double price
They'd richly own the same sir

New Bedford folks are fond of whales
The largest are the best sir

And when the sperm whale cannot be had
Then right whales they will do sir

Indeed a very few days since
Coffin he will go sir
We wish him a good passage home
To see his friends quite well sir

And when he's on the deep blue sea
Of him we'll gladly think sir
And wish him every blessing home
All this he well deserves sir

Maria (2) 1846

Both this song and "A Whaling Scene" were put in the *Maria*
journal at the end of the voyage when the vessel was lying at St. Helena
full of oil and homeward bound. The news of the gold strike in California had even then reached that remote island.

THE WOUNDED WHALE

See the sun from her ocean bed rises
Broad o'er the water glittering light throws
But hark from out masthead what news are you bringing
Hard on our lee beam a whale there she blows

SONGS THE WHALEMEN SANG

Call up your sleepers both your larboard and starboard men
Main yard aback and your boats on your lee beam
The white waters are gleaming
Gleaming and foaming in gallant array

See the living then in vastness is lying
Making the sea her luxurious bed
While high in the air the sea birds are playing
Combing the billows that breaks o'er her head

Whilst high wide and sidewise goes her dark flukes in the air
Stately but slowly she sinks in the main
Peak all your oars awhile rest from your weary toil
Waiting and watching her rising again

Row hearties row for the pride of your nation
Spring to your oars make the reeking sweat flow
Now for the blood let it have circulation
Foreward from your thwart give way every man

See how the boats advance gaily as to a dance
Floating like shadows across the blue sea strand
Up and give him some send both your irons home
Safely stern all trim the boat all clear he's wounded

Wounded and sore fins and flukes in commotion
Blackskin and oars contending in the spray
While so loud and so shrill blows the horn of the ocean
Fretted and lost she brings to in dismay

Haul line every man gather in all you can
Lances and spades from your thwarts clear away
Now take your oars again fast each boat (removes)
Waiting and watching her rising again

Surrounded by foes with strength undiminished
Heed now how she flashes her dark flukes in the air

24

A lance in the life and the struggle is finished
Oh oh she sinks with her chimney on fire

While so loud and so shrill the cry from our seamen
Mocking the whale in her terrible (hour)
Now looking at her die see the blue signal fly
Here she goes fin out and the contest is o'er

Dartmouth 1836

THE WOUNDED WHALE
(Second Version)

Lo as the sun from her ocean bed was rising
Broad on the ocean its glittering light threw
Hark from the mastheads our lookouts are crying
'Tis hard on our lee beam a whale there she blows

Call up your sleepers larboard and starboard men
Main yard aback and the boats clear away
For 'tis hard on our lee beam see the white water gleam
Glittering and foaming in gorgeous array

See the old leviathan fretting is lying
Making the ocean her sumptuous bed
High in the air the sea birds are flying
Combing each billow that breaks o'er her head

High wide and swimming dark waves are flying
Slowly but surely he sinks in the main
Now peak your oars awhile
Waiting and watching her rising again

Row hearties row for the pride of your nation
Spring to your oars let the raging sweat flow
And now for the blood let it have circulation
Foreward on your thwarts give way all you know

See how the boat advances gaily as to a dance
Floating like a feather o'er the dark blue sea
Stand up and give him some send both your irons home
Stern off trim your boat we are all clear

Wounded and sore a thousand times commotion
Paddles and oars contend with the spray
Loud and so shrill the horn of the ocean
Wounded and fatigued she brings to in dismay

Haul line every man gather in all you can
Lances and spades from your thwarts clear away
Now peak your oars again while fast each boat remains
For safely and surely we hold him at bay

Wounded and sore yet strength undiminished
He lashes the sea in his ire
A lance in the life and the struggle is finished
As he sinks down with his chimney on fire

Loud rings the shouts from every seaman's mouth
Matching the sea in its turbulent roar
Now see the whale dies let your red signal fly
For safely and sure the contest is o'er

Uncas 1843

This song must have been very popular with whalemen despite the fact that at its conception it must also have been decidedly literary — or I should say flowery? Both the *Dartmouth and Uncas* versions are garbled in spots, but the tremendous excitement of going on and killing the whale is in both of them. There is a very good version of this song in Joanna Colcord's *Songs of American Sailormen*, pp. 189-190. She calls it "There She Blows."

"Send both your irons home." The iron was the harpoon. So this line means to put two harpoons in the whale to make doubly sure of him. It wasn't always, by a long way, that the boatsteerer had time to do that.

"Stern all" means back the boat away from the whale after the irons were placed, to be clear of his flukes. That "stern all" was a most welcome order to the rowers who were never able to see half of what was going on, for it meant that now, temporarily at least, they were out of danger and fast to the whale.

"See the blue signal fly." Here the blue signal meant that the whale was dead. In the *Uncas* version it is "Let your red signal fly."

ROLLING DOWN TO OLD MOHEE

Once more we are waft by the northern gales
Bounding over the main
And now the hills of the tropic isles
We soon shall see again
Five sluggish moons have waxed and waned
Since from the shore sailed we

SONGS THE WHALEMEN SANG

Now we are bound from the Arctic ground
Rolling down to old Mohee
Now we are bound from the Arctic ground
Rolling down to old Mohee

Through many a blow of frost and snow
And bitter squalls of hail
Our spars were bent and our canvas rent
As we braved the northern gale
The horrid isles of ice cut tiles
That deck the Arctic sea
Are many many leagues astern
As we sail to old Mohee
Are many many leagues astern
As we sail to old Mohee

Through many a gale of snow and hail
Our good ship bore away
And in the midst of the moonbeam's kiss
We slept in St. Lawrence Bay
And many a day we whiled away
In the bold Kamchatka Sea
And we'll think of that as we laugh and chat
With the girls of old Mohee
And we'll think of that as we laugh and chat
With the girls of old Mohee

An ample share of toil and care
We whalmen undergo
But when it's over what care we
How the bitter blast may blow
We are homeward bound that joyful sound
And yet it may not be
But we'll think of that as we laugh and chat
With the girls of old Mohee
But we'll think of that as we laugh and chat
With the girls of old Mohee

Atkins Adams 1858

28

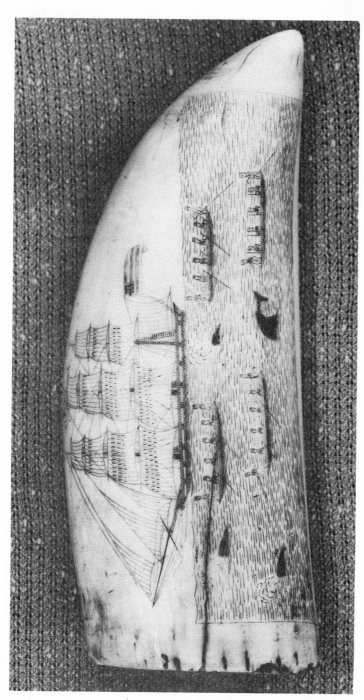

Scrimshaw on sperm whale's tooth
And if you get too near his flukes
He'll kick you to the devil

. Joanna Colcord says that she couldn't find the
. But Harlow, pp. 228-229 does have a melody
I am using here. It almost fits the *Atkins Adams*
after a little tailoring.

Memory will play strange tricks over a matter of fifty years or
so, but I remember Bill Tilton singing this song, and it seems to me that
the melody he used was "The Bowery." It fits, but I didn't have the
temerity to use it here. Sigmund Spaeth in *Read 'Em and Weep,* p. 186,
notes that "The Bowery" is perhaps related to the Italian street song,
"Spagnola." Could it be that both "The Bowery" and Bill Tilton's tune
for "Rolling Down to Old Mohee" derive from "Spagnola?"

"We are homeward bound that joyful sound and yet it may not
be," must refer to the fact that the oil from a full whaler was often
transshipped home in another vessel and so the whaler wasn't "homeward
bound" after all, but put to sea again to seek more whales.

DIEGO'S BOLD SHORE

SONGS OF WHALES AND WHALING

Has a love of adventure a promise of gold
Or an ardent desire to roam
Ever tempted you far o'er the watery world
Away from your kindred and home
With a storm beaten captain free hearted and bold
And a score of brave fellows or two
Inured to the hardship of hunger and cold
A fearless and jolly good crew

Have you ever stood watch where Diego's bold shore
Looms up from the Antarctic wave
Where the snowy plumed albatross merrily soars
O'er many a mariner's grave
Have you heard the masthead'sman sing out there she blows
Seen the boats gaily leave the ship's side
Or the giant fish breach 'neath the harpooner's blows
Till the blue sea with crimson was died

Have you seen the foam fly when the mighty right whale
Thus boldly attacked in his lair
With a terrible blow of his ponderous tail
Sent the boat spinning up in the air
Or where the green isles of the evergreen glades
Are teeming with dainties so rare
Have you ever made love neath the coco's green shade
To the sweet sunny maids that dwell there

Let those who delight in the comforts of home
And the joys of a warm fireside
Who dream it a peril the ocean to roam
In the cots of their fathers abide
But not a day nearer we reckon our death
Though daily we sport o'er our graves
Nor sweeter they'll slumber beneath the green sod
Than we in the boisterous waves

Have you ever joined in the boisterous shout
Reaching far through the heaven's blue dome

31

When rich in the spoils you have purchased so dear
You have hoisted your topsails for home
Or when the dark hills of Columbia arose
From out the blue waves of the main
Have you ever relived the unspeakable joy
Of meeting with loved ones again

Midas 1861

This seems to have been one of the best loved of the traditional whalemen's songs. Joanna Colcord in *Songs of American Sailormen* has a version very similar to this except that it lacks the fourth stanza. It is her melody that I have used here.

THE OLD HULK

When age has rendered some old hulk
Unfit for merchant use
She's sold at auction bought in bulk
Just for a whaling cruise

Now paint and tar renews her age
A-1 once more stands she
The agents then a crew engage
Scarce half ne'er saw the sea

With casks in shooks and beef and pork
The ground tier is chocked and stowed
The last for sailors' jaws to work
The first to hold the load

With boats and tubs and lines prepared
The whaler's underway
To cruise where e'er a ship has dared
A floating Botany Bay

Around the stormy southern cape
A fearless course we steer

Then northward where the storm clouds drape
The sky throughout the year

Our cruising ground is gained and then
Aloft loud rings the hail
She blows look out to windward men
A ninety barrel whale

Down go the boats he first who can
On for the prize we dash
The hunt is up and every man
Bends to the buckling ash

Quick lay me on hurrah we're fast
Stern all lay off my hearts
The monster's life is reached at last
With lance and barbed darts

Again long luckless months go by
And not a whale is found
And we for better fortune try
Some other cruising ground

Three years have sped towards home once more
Scarcely half full of oil
What reck we so we gain the shore
And tread our native soil

Our voyage is up and for our lay
We'll take what we can get
But find instead of getting pay
We're fifty cents in debt

Governor Carver 1854

In spite of the fact that the song says otherwise, most whaleships were sound and seaworthy. They had to be. The song is nearer right in the line about "A floating Botany Bay." For some whalers were pretty nearly prison ships.

The shooks in the third stanza were the staves of casks and hogsheads tied up in bundles. It was the cooper's job to put the barrels and casks together as needed.

THE BARK GAY HEAD

Come all you young Americans
And listen to my ditty
It's all about a whaling bark
That left New Bedford city
The bark Gay Head it is her name
She's known both far and near

Her rules and regulations
They are most awful queer

Chorus

Cheer boys cheer for the Gay Head and crew
For growling and soldiering when there are things to do
We never will we never will we never can be high
We want three thousand barrels of oil root hog or die

The place this noble bark was built
Was Mattapoisett town
The workmen worked for a dollar a day
The boss was Jonathan Brown
She was launched upon July the tenth
At eight o'clock P.M.
And numerous crowds assembled round
To see them dump her down

The Captain's name was Jenny
From Mattapoisett town
He walks upon the quarter-deck
And there you'll see him frown
He is the meanest captain
That ever you did see
He's crossed the Atlantic several times
From New Bedford to Africee

We'll cheer my noble hearties
For the larboard boat and crew
Mr. Hazzard's their boat leader
He's a gentleman good and true
There's Hussy John and Taylor Dick
And a boatsteerer named Couch
And when they lower in their boat
They know what they're about

And we will cheer my hearties
For the starboard boat and crew
For the blowing of their pulling
Which is something they can do
When they lower in their boat
They do the best they can
Ryder he's the biggest blower
And he's the smallest man

We have the noble waist boat
Whose crew are very good
Their boatheader is a Dartmouth man
His name is Mr. Wood
And when they lower in their boat
They don't make any noise
But when you talk of pulling
They are the very boys

We have another boat now
It's called the bow boat
She's one of the jolliest crafts
That ever yet did float
Hiller pulls the bow oar
And Blankenship the stroke
Dexter pulls harpooner
And Jenny heads the boat

Stella 1860

This documentary song sings to the tune of "Root Hog or Die." It is interesting as it identifies the members of the various boats' crews. Almost all the names of these men are still to be found in the Buzzards Bay area.

There were two different whaling vessels named the *Gay Head*. I think that this must be the earlier one. Mattapoisett was noted for the whale ships that were built there.

THE BARK OCEAN ROVER

Many and trim are the whalers that appear
A-cruising the New Holland ground over
But of all that is there there is none to compare
With the neat little bark Ocean Rover

Chorus

Oh merrily merrily goes our bark
Before the gale she bounds
So flies the dolphin from the shark
Or the dear before the hounds

Her movements are graceful as those of a doe
She's as fleet as a dove when in motion
And she is acknowledged by all on the ground
To be the pride of the Indian Ocean

We have tried them all under close-reefed mainsail
And under top-gallant sails too
But ha-ha they all cry the whalers are many
But those that can beat us are few

There's the Pamelia they blow on her sailing
They say she can never be beat
But whenever the Ocean Rover is 'round
It's then she is done up so neat

It's not long since she was running to catch us
With her main top-gallant sail out
With her mizzen staysail fly-jib and gaff topsail
But after all we had to veer about

We first took in the main then the topsail hauled a-back
And then the jib we hauled down
She (still) couldn't catch us so the (ousail) we hauled up
I'll be buggered if the Rover isn't sound

There may be some can beat us but they mustn't be slow
And if they can beat us why then they can blow
She is named the Rover for she is always 'round
Where ever there is whalers to be found

Now to finish my song which is very long
She soon will be homeward bound
And if you should gam her just bear it in mind
That the Rover is always around

Ocean Rover 1859

In "The Bark Gay Head" we see the rivalry between the boats' crews, while this song tells of the rivalry between the vessels.

So far as I can find out, this particular bark *Ocean Rover* made only one voyage whaling and then was sold to the Spanish government, probably to be used as a dispatch boat. But there was another bark *Ocean Rover* which was captured and burned by the Confederate cruiser *Alabama* in 1862.

The gam in the last stanza was a visit at sea between the crews of two or more whaleships. The vessels would be hove to, and the crews would visit back and forth. It was a time to meet old friends and to exchange mail as well as news and gossip. Songs were exchanged, too, during gams.

DESOLATION

I will sing a little rhyme as I have a little time
About the meanest ship afloat in creation
Her name it is the Mitchell from E town did sail
And they fitted her out to go to Desolation

Her officers are natives of old Cape Cod
The place where there is nothing to eat on
But the product of their land is mackerel bones and sand
So they had to starve or go to Desolation

On board of some ships they have plenty to eat
But it is here they put a stop on our ration

38

It is work for nothing and find your own grub
And starve yourself to death on Desolation

The meat on this ship once belonged to a horse
Or some of his damned near relation
They put us on an allowance of a quarter of a pound
They could afford no more on Desolation

For fear the flour would not last for bread three times a day
And mince pies to feed the after guard on
They cut us short one half and says with a laugh
It's good enough for Jack on Desolation

The captains of whalers are abolitionists
They go in for amalgamation
A nigger or a Portuguese is treated like a man
But Americans are dogs on Desolation

These cowards and villains for they are such a race
They are a disgrace to all civilization
Are our worthy friends who call themselves men
And command these prison hulks on Desolation

For toward the end of the voyage they treat you mighty rough
They cause you trials and tribulations
For if you have any pay they would have you run away
And pocket all your earnings on Desolation

Ocean Rover 1859

This is the only version of this song that I have seen, but it may have had some currency for it was recorded in the *Ocean Rover* journal.

"E town" must be Edgartown. I have never heard Vineyarders use that expression, but many Vineyarders do refer to Provincetown as "P town." Very, very rude remarks were often made by Vineyarders about the good people of Cape Cod, compared to which that about "The product of their land was mackerel bones and sand," is mild and kind.

The stanza about the captains of whalers being abolitionists stresses the fact that there was so often friction in the forecastle between the Yankees in the crew and the Portuguese who very often were colored and from the Cape Verde Islands. The Indians in the crew were not considered colored at all, but only slightly darker-skinned Yankees, and they, too, all too often, shared the feeling against the "Portugees." I have never seen any mention of a similar feeling against the Kanakas who often served as crew members in the Pacific.

The idea expressed in the last stanza is true enough, for there were vessels where an attempt would be made to get crew members who had a good lay, or share, coming to them to desert. The persuasion took the form of sheer brutality, and if they did skip ship there would be more money to divide up among the members of the afterguard and the owners.

THE WINGS OF A GONEY

If I had the wings of a goney
I would fly to my own native home
I would leave Desolation's cold weary ground
For for right whales for us there is none

For the weather is rough and the wind it does blow
And comforts are not to be found here
I would sooner be at home in some Dutch grocery shop
Eating crackers and cheese or drinking beer

For a man must be foolish to venture so far
On the broad blue expanse catching whales
When he knows that his life is in danger at times
Or his head being smashed by their tails

But whaling has its charms for the young and green hands
And he makes up his mind when he goes
In a very short time he would sooner hear a curse
As the unwelcome sound there she blows

SONGS OF WHALES AND WHALING

It is first learn the compass with thirty-two points
Or else lose your whole watch below
But that is not all to his sorrow he will find
Two hours to the masthead he must go

He descends to the deck with head dizzy and sick
And for life he would not give a damn
When the mate accosts him with Johnny my boy
Get a rubber and rub down spun yarn

After bearing these trials for well nigh four years
The flying jibboom points for home
He's supposed to have eight or ten barrels of oil
And an equal share of the bone

But he goes to the agents to settle his voyage
And there he finds cause to repent
He finds he has spent four years of his life
And not earned a single red cent

Ocean Rover 1859

The goney is a bird of the albatross family. Today American sailors in the Pacific call it the "gooney bird."

The last two stanzas of the song tell a story that, sad to tell, was often only too true. Whalemen could become deeply in debt to the ship for clothing and tobacco from the slop chest, and perhaps also, sometimes for advances for shore leave. And thus at the end of a three or four year voyage, the whaleman might have nothing at all to show for those years out of his life. It was to avoid this danger that many whalemen wore clothing that had been patched and repatched and then patched again.

This song sings very well to the tune of "The Prisoner's Song," which begins, "If I had the wings of an eagle across the wide sea I would fly." I wonder if there can be any connection between the two songs.

To show how a sailor would become indebted to the vessel, here is part of a page from the *Elizabeth* journal (1847).

Thomas R. Bryant to Ship Elizabeth, Dr.

Feb. 1, 1849	for two lbs. tobacco	.60
Feb. 7	for one lb. tobacco	.30
March 25	4 yds. cotton cloth	1.00
March 29	4 yds. cotton cloth	1.00
June 5	1 jack knife	.30
July 6	2 fathoms blue cloth	1.20
July 7	2 lbs. tobacco	.60
July 8	cash	1.00

And a bill like this over a period of years would add up.

BLOW YE WINDS

'Tis advertised in Boston
New York and Albany
Five hundred young Americans
Are wanted for the sea

They take you down to Bedford
That famous whaling port
And give you to some landsharks
To board and fit you out

It's then that they will show you
Their fine clipper ships
They say you'll have five hundred sperm
Before you're six months out

It's now we're out to sea my boys
The wind comes on to blow
One half the watch is sick on deck
The other half sick below

Next comes that damned old compass
It will grieve your heart full sore
For theirs is two and thirty points
And we have forty-four

Next comes the running rigging
That all of you must know
And if you don't know it in fifteen days out
You'll lose your watch below

It's now to the masthead
All of us must go
And when you see those sperm whales
Sing out there she blows

Then up will step the skipper
And sing out where so
Where away you damn landlubber
Does that sperm whale blow

Then tell him with a cheerful voice
Three points on our lee bow
I think he's going to leeward
My boys going very slow

Then clear away the boats my boys
And after him we'll go (travel)

Whale stamps from journal page (each stamp in the ship's
log meant a dead whale)
Bowheads and blackskins

And if you get too near his flukes
He'll kick you to the devil

Now we have got him turned up-side (down)
We'll tow him alongside
And over with our blubber hooks
To rob him of his hide

Now the boatsteerer overboard
The tackle overhauled
The skipper's in the main chain
So loudly does he bawl

Next comes the stowing down my boys
It will take both night and day
And you'll all have fifty cents apiece
On the hundred and ninetieth lay

Now we are all tired out
With sailing all about
For blackskin breeches sporting
And watching for him spout

O portward bound it is the sound
With a stock of education
Where liberty men both now and then
Bring trouble and vexation

Now we are bound into Tombas
That damned old whaling port
And if you run away my boys
You surely will be caught

Now we're bound into Tuckoona
Full more in their power
Where the skipper can buy the council up
For a half a barrel of flour

It's now that our old ship is full
And we don't give a damn
We'll bend on all our stuncils
And sail for Yankee land

And when we arrive in Bedford
Whaling we bid adieu
And if any of you fellows a flogging have got
Just put your skipper through

Here's to all skippers and all mates
I wish you may all do well
And when you die may the devil
Kick you all into hell

Now we got home our ship made fast
And we got through our sailing
A winding glass around we'll pass
And damn this blubber whaling

Elizabeth Swift 1859

"Blow Ye Winds" is one of the best, as well as one of the best known of the whaling songs. This version from the *Elizabeth Swift* journal is called only "Whaling Song" and there is no hint or indication of a chorus. But the song always was sung with a chorus, and indeed, was often used as a chantey.

The compass with forty-four points, as Joanna Colcord points out in *Songs of American Sailormen* is a mystery. There must have been such a compass, but how in the world do you fit eleven points into ninety degrees, and what were those points? The standard mariner's compass has thirty-two points.

See Colcord, pp. 190-193; Hugil, pp. 219-224; and Harlow, pp. 130-131.

Of The Sea And Ships

HERE ARE two quite different kinds of sea songs; those that sailors themselves made like "The Ship Euphrasia" and "Hearts of Gold," and the art songs like "A Wet Sheet and a Flowing Sea" and "The Topsail Shivers in the Wind." But if an art song dealing with the sea was right and true in its spirit and terminology, seamen felt no qualms in accepting it.

Some of the beauty and mystery of the sea is here as well as much of its danger. As the whalemen were seamen as well as the practitioners of a highly specialized craft, the songs of the sea belonged to them as well as to merchant sailors.

THE SHIP EUPHRASIA

Come all Christian people who do intend
To know God's laws and his rights defend
Just think for a moment on my cruel fate
That on the wide ocean to you I'll relate

On the twelfth of November our canvas was spread
And our noble ship was soon hove ahead
With the bright stars and stripes at her mizzen peak
Kind friends and relations we bid all (adieu)

SONGS THE WHALEMEN SANG

But little I thought what I should go under (undergo)
So long on the ocean where stormy winds blow
For me being a stranger unto the salt sea
I thought of no trouble but we could agree

Until on board kept on such filthy fare
It has often times caused me to stamp and to swear
For we have nothing fit for a Christian to eat
For most of the time it is old stinking meat

Sometimes meal and maggots mixed up in a tub
One of our great messes some think it good
With the sweat that ran down the old negro's face
All mixed up together Oh Lord what a taste

We have some potatoes half rotten to eat
Mixed up by the negro with poor stinking meat
Not fit for a hog without (rine) like a boar
And after one eating would never want more

Our noble commander 'tis seldom we see
He sticks to his hive the same as a bee
Contriving some plan I suppose for to beat
From us everything wholesome and fit for to eat

The most of our living on board that we had
Would make all the dogs in creation run mad
And still I believe in a God ever true
And our big glutted hero will still get his due

For when he is gone from all things here below
To the torments of hell then I think he will go
Where all the young devils will laugh at the fun
And feed him up well on Brimstone and rum.

No good gin and brandy they'll (wine) him I hope
No more good old Spanish will he ever smoke

But a common long nine to keep him awake
While he's rolling on hell's quarter-deck

If I was a painter I vow and declare
I would draw you a picture of our dreadful fare
You never would wonder why I did complain
Or such cruel usage all on the ruff main

My song now is ended I'll write you no more
Until I have landed on that golden shore
When I am resolved to tell unto you
What our big glutted hero has tried for to do

Euphrasia 1849

Perhaps this song was sung nowhere else but on board the ship *Euphrasia*. And probably it was not sung too loudly there. It is crude and poor in places but I think it should be included here, for constant complaining about the food was a part of life on many vessels. A note in the journal says that this song was composed by Isack Bray of Rockport.

A WET SHEET AND A FLOWING SEA

A wet sheet and a flowing sea
And a wind that follows fast

That fills the white and rustling sail
And bends the gallant mast
That bends the gallant mast my boys
While like an eagle free
Away our good ship flies and leaves
Columbia on our lee

Oh for a soft and gentle wind
I heard a fair one cry
But give to me the roaring breeze
And the white waves heaving high
The white waves heaving high my boys
And a good ship tight and free
The world of waters is our home
And merry men are we

There's a tempest in yon horned moon
And lightning in yon cloud
And hark the music mariners
The wind is piping loud
The wind is piping loud my boys
The lightning flashes free
While the hollow oak our palace is
Our heritage the sea

Citizen 1844
Cortes 1847

Like "A Life on the Ocean Wave" this is a literary song that sailors took over and made their own. In the original it was "Britannia on Our Lee." That as always has been changed to "Columbia."

The *Cortes* and *Citizen* versions are very similar as one would expect with a song of this type. The song was composed by Allan Cunningham, a Scottish poet who was also a collector of folk songs. And the melody seems to be a very old French air.

SLING THE FLOWING BOWL

Come come my jolly lads the wind's abaft
Brisk gales our sails shall crowd
Come bustle bustle bustle boys haul the boat
The boatswain pipes aloud
The ship's unmoored all hands on board
The rising gale fills every sail
The ship's well manned and stored

Chorus

Then sling the flowing bowl
Fond hopes arise the girls we prize

Shall bless each jovial soul
The can boys bring we'll drink and sing
While foaming billows roll

Though to the Spanish coast we're bound to steer
Our rights we'll there maintain
Then bear a hand be steady boys
Soon we'll see New England once again
From shore to shore while cannons roar
Our tars shall show the haughty foe
Columbia rules the main

Chile   1839
Cortes 1847

This song seems to have become pretty well standardized as both the *Chile* and *Cortes* versions as well as all printed versions that I have seen are quite similar. But the melody does show change. The one I have used here is from Kitchiner, pp. 58-59, and it is quite different from the one in Whall, pp. 17-19.

As one would expect, Old England has been changed to New England, and Britannia to Columbia in the journal versions.

LOOSE EVERY SAIL TO THE BREEZE

Loose every sail to the breeze
The course of my vessel improve
I've done with the strife of the seas
Ye sailors I'm bound to my love

Chorus
Ye sailors I'm bound to my love
Ye sailors I'm bound to my love
I've done with the strife of the seas
Ye sailors I'm bound to my love

Since Emma is true as she's fair
My griefs I shall fling all to the wind
'Tis a pleasing return for my care
Where no conflict but love shall I find

My sails are all filled to my dear
What tropic bird faster can move
Who cruel shall hold his career
That returns to the nest of his love

Hoist every sail to the wind
Come shipmates and join in the song
Let's drink while our ship sails the seas
To the gale that may drive her along

Joseph Francis 1795

In the journal this fine old song is called "Homeward Bound." But there are so many homeward bound songs that I have given it the title it has in Kitchiner's *Sea Songs of England.* In the journal there is no chorus, but I believe the chorus was sung. See Whall, pp. 35-37, and *The American Musical Miscellany*, pp. 202-204.

Note the terrific range of two full octaves for this song.

CAPTAIN JAMES

Captains listen to my story
A warning you must take by me
See that you don't abuse your sailors
While you're rolling on the sea

Richard Paddy was my servant
A handsome sprightly lad was he
His mother bound him to me an apprentice
For to cross the rolling sea

As we had been to South Carolina
And we were returning home
Cruelly this boy I murdered
Such a thing was never known

A trifling offense he gave me
Which did my loving heart enrage
Straightway to the mast I bound him
There I kept him several days

With his hand and arms extended
I no succor to him gave
Swearing if my men relieved him
Not a moment should he live

When three days I there had kept him
He with hunger loud did cry

OF THE SEA AND SHIPS

Now for God's sake pray relieve me
Or with hunger I shall die

Eighteen bitter stripes I gave him
Which did cause the purple gore to run
None there was that dare relieve him
Such a thing was never known

When five days I thus had kept him
He to languish did begin
Praying for a little water
I some vinegar gave to him

The poor soul requested to drink it
As I had proposed when I had done
I made him drink the purple gore
That from his bleeding wound did run

When many days I thus had kept him
Up to him I then did go
He says my dearest loving master
One small favor to me show

Don't leave me here thus for to suffer
Kill and send me to the grave
One small piece of bread afford me
Which in humanity I crave

Oh that I had but one small morsel
Which the dogs they do despise
He says oh Lord send me some water
From the lofty blissful skies

Hearing what he said unto me
Would have grieved a Christian's heart
Often time he cried dear mother
Did you but know the cruel smart

How your tender son does suffer
It would grieve you to the heart
More bitter grief no tongue can utter
Lord relieve me from my fate

When my men they disobeyed me
I like a fury cursed and swore
That I would have them hang for pirates
When I did arrive on shore

But they knowing my intention
Little to me they did say
And they had me apprehended
When I had got home from sea

How can I now ask for mercy
When no mercy I would afford
On a poor distressed creature
Yet some mercy show me Lord

I thought that my money would have saved me
Knowing that the boy was poor
But the cries of his dear mother
Would have grieved a heart full sore

She was resolved to prosecute me
She no gold nor bribe would take
Captain James for cruel murder
Now the gibbet is your fate

Walter Scott 1840

CAPTAIN JAMES
(Second Version)

Come all you noble bold commanders
That of the raging sea do use
By my sad fate pray take a warning
That your poor sailors you don't abuse

OF THE SEA AND SHIPS

Richard Peve he was my servant
And a sprightly boy was he
His parents did apprentice bind him
For to cruise the raging sea

'Twas on a voyage to Carolina
As we were returning home
Cruelly this boy I murdered
Such a thing was never heard

'Twas some little offense he gave me
Which set my bloody heart in rage
Straightway to the masthead I sent him
Where I kept him many a day

With his legs and hands extended
Him no succor did I give
Saying if my men did relieve him
Not one moment would they live

When three days I thus had kept him
Up to him I did go
He did cry so dear and loving
. . . . . .

Cortes 1847

CAPTAIN JAMES (Third Version)

Come all you noble brave commanders
Come and a warning take by me
See that you don't abuse your seamen
When you are on the raging sea

Richard Spry he was my prentice
And a likely lad was he
His parents bid me prentice bind him
For to cross the raging sea

SONGS THE WHALEMEN SANG

It was on a voyage to Carolina
In our passage coming home
Cruelly I this boy did murder
Which was a fact heaven n'er known

It was for some trifling acts he gave
Which put my bloody heart in rage
To the mast I straightway tied him
There I kept him many days

All my men they did (upbraid) me
But bitterly I stomped and swore
That I would have them all hung for piracy
If ever to England we got o'er

When nine days I there had kept him
Loud for hunger he did cry
Praying to Jesus to send him water
From the high and blessed sky

Meat and drink I then did bring him
Utterly he did refuse
Bitter stripes I then did give him
And his (body) did sorely bruise

Do not keep me here forever
Kill me send me to my grave
Or else some mouldy crusts afford me
That the dogs will refuse to have

I thought my money would have favored me
Knowing his parents to be poor
Loud cries his tender mother
Makes me grieve my heart full sore

I have got gold out of measure
And silver laid up in store

OF THE SEA AND SHIPS

Now captain James for cruel murder
Hang in gibbits by a lee shore

Two Brothers 1768

This song must tell the story of an actual crime for which Captain James was hanged. The long *Walter Scott* version looks as though it may follow a broadside pretty closely. But the other two do not. I would hazard the guess that it had very wide currency.

The fragment in the *Cortes* manuscript is not in Histed's handwriting, and the spelling is terrible (Histed's spelling is uniformly good and he must have been a man of some education) but I do wish that whoever put this song in Histed's book had finished the story. The *Two Brothers* version is very early indeed, and so we know that the song is old.

There are versions of the song in JFSS, vol. 2, pp. 161-162, and vol. 7, pp. 4-5 and 66-67. The notes for this song are interesting. In each case it is called "The Captain's Apprentice."

THE TOPSAIL SHIVERS IN THE WIND

The topsail shivers in the wind
Our ship she casts to sea
But yet my soul my heart my mind
Are Mary moored with thee

59

For though thy sailor's bound afar
Still love shall be my leading star

Should landsmen flatter when we're sailed
Oh doubt their artful tale
No gallant sailor ever failed
If love breathed constant gales
And love shall steer my heart and soul
Shall steer my heart from pole to pole

These cares are ours but if you're kind
We'll scorn the dashing main
The rocks and billows and the wind
Till we return again
Columbia's glory rests with you
Our sails are full sweet girl aideu

Ann 1776
Frances Henrietta 1835

This song goes back well into the eighteenth century and although its origins were literary it is authentic enough both in spirit and terminology so that seamen took it over. Probably it should be sung slowly and with pauses, for it seems to be of the "come-all-ye" type.

There is an extra stanza to the song from the version in the *Ann* journal.

Sirens we find in every port
More fatal than the rocks and waves
But such as grace our British fleets
Are lovers and not slaves
No face can ever us subdue
Although we leave our hearts with you.

Also, in the last stanza of the *Ann* version it is "New England's glory" rather than Columbia's. For British versions of the song see Stone, p. 189, where it's called "The Sailor's Adieu," and Kitchiner, pp. 60-61.

In the South Atlantic Ocean. 1838.

Sunday Nov 25th 1838.
Commences with light winds from the Northward
saw plenty of whales lowered and chased starboard boat
to oil and he stove the boat and were obliged to cut the
line and let him go in safety and latter part light
breezes saw whales and chased waist boat struck and drawed.
So ends these 24 hours.   Lat. 36. 08.
                            Long. 13. 59.

W. B.   T. B.

Monday Nov 26th 1838.
Begins with very light breezes and chased a right
whale, middle part nearly calm latter part fine breezes
from the Northward saw whales and chased without
success. and caught two porpoises.
        So ends these 24 hours.   Lat. by obs. 36. 32.
                                    Long. by ct. 24. 25.
missed

Tuesday Nov 27th 1838.
Commences with fresh breezes from the Northward
saw plenty of whales and chased middle part fresh
breezes latter part strong breezes saw right whales
and chased.
        So ends these 24 hours. Lat. by D.R. 36. 50.

Wednesday. Nov 28th 1838.
Begins with light breezes from the Westward and
fine weather saw right whales and chased but did
not strike. middle part much the same latter part
fresh breezes saw a right whale and chased.
        So ends these 24 hours.   Lat. by obs. 36. 43
                                    Long. by ct. 24. 07.
missed   W. B.

Page from a journal
    We have the noble waist boat
    Whose crew are very good

SONGS THE WHALEMEN SANG

THE SEQUEL TO WILL WATCH

'Twas the girl the girl that Will Watch loved dearly
Heaved a sigh and turned pale when she heard of his death
For ne'er was affection returned more sincerely
Than that by his Susan while Susan had breath

Brave Will prized her merits far more than her beauty
Though Susan was lovely as lovely could be
But merit with Will was a jewel and duty
To love and to fight for at home and at sea

'Twas her hand tied his handkerchief when they last parted
'Twas her bosom pressed his as they stood on the beach
'Twas his lips that kissed off the fond tear that started
And did for his Susan each blessing beseech

Will swore naught in life his attachment could sever
His heart was his Susan's by land and by sea
Yet should it so happen we now part for ever
Then wed some good fellow and love him for me

He spoke fled and fought aye and died like a man too
For Will was soon cut off at destiny's call
Yet the boast of his crew is and truly they can too
How dearly Will Watch was beloved by them all

The news of his fate with reluctance and sorrow
The very next day to his Susan they bore
She heard and frenzy her wits seemed to borrow
She smiled looked around her and never spoke more

In the grave with the lad she both lived and died for
Were laid the remains of the girl he loved dear
And while to his memory his mates heaved a sigh for
Each lover will give to his Susan a tear

Not a flint marks the spot where their bones lie enshrouded
Yet the earth is held sacred and dear by the crew
And often right oft by the moonbeams enclouded
Is a tear dropped for Will and his Susan so (true)

Cortes 1847

"Will Watch" as Whall points out in *Sea Songs and Shanties,* p. 39, was popular with sailors because it was at one and the same time both very professional and very sentimental. This sequel to "Will Watch" may not be overly professional but it surely is sentimental. Whall thinks that the parent song was composed about 1820.

THE SEA

The sea the sea the open sea
The blue the fresh the ever free
Without a mark without a bound
It runneth the earth's wide regions round
It plays with the clouds and mocks the skys
Or like a cradled creature lies

I love oh how I love to ride
On the fierce foaming bursting tide
Where every mad wave drowns the moon
And whistles aloud its tempest tune

And tells how goeth the world below
Or why the southwest wind does blow

I'm on the sea I'm on the sea
I am where that I would ever be
With the blue above and the blue below
And white waves where so ever I go
If storms should arise and wake the deep
What matter I can ride and sleep

I never was on the dull tame shore
But what I loved the great sea more
And backward flew to its billowy breast
Like a bird that seeks its mother's nest
For a mother it was and is to me
For I was born on the deep blue sea

The waves were white and red the moon
In the stormy hour when I was born
And the whale it whisked and the porpoise rolled
And the dolphin bared its back of gold
And never was heard such an outcry wild
As welcomed the life of the ocean child

Since then I've lived in calm and strife
Full fifty summers a seaman's life
With wealth to spend and power to range
And never sought nor sighed for change
And death when ever it comes for me
Must come on the deep and boundless sea

Cortes 1847

There is a version of this song in H. K. Johnson's *Our Familiar Songs* with an involved musical setting. The melody as I have it here is greatly simplified.

OF THE SEA AND SHIPS

SATURDAY NIGHT AT SEA

A sailor loves a gallant ship
And messmates bold and free
And ever welcomes with delight
Saturday night at sea
Saturday night at sea my boys
Saturday night at sea
Let every gallant sailor sing
Saturday night at sea

One hour each week we'll snatch from care
As through the world we roam
And think of dear ones far away
And all the joys of home
Saturday night at sea my boys
Saturday night at sea
Let the winds blow high on board
Saturday night at sea

We'll think of those bright beings who
Bedeck with joy our lives

And raise to heaven a prayer to bless
Our sweethearts and our wives
Saturday night at sea my boys
Saturday night at sea
In storms and calms through life we'll sing
Saturday night at sea

Florida 1843

There is a version of this song in *The American Musical Miscellany*, pp. 120-122. Also, here is an extra stanza from a version in the *Mammoth Songster* which was published in Boston in 1866:

Come messmates fill the cheerful bowl
Tonight let no one fail
No matter how the billows roll
Or roars the ocean gale
There's toil and danger in our lives
But let us jovial be
And drink to sweethearts and to wives
On Saturday night at sea.

So if the weather was fair, Saturday night meant a little while on deck to sing and relax, and depending on the skipper, perhaps there was even an issue of grog.

I WAS ONCE A SAILOR

Yes I was once a sailor lad
I plowed the restless sea

OF THE SEA AND SHIPS

I saw the sky look fair and glad
And I felt proud and free

I breathed the air of many a clime
Saw beauties fair and gay
My hopes were fixed on future time
The present slipped away

Experience sad hope's brilliant view
Like mist dissolved away
I found small harvest did accrue
To plowmen of the sea

I found my team would rage and rove
'Twas but the fickle wind
That plowing o'er the rolling sea
No furrow left behind

Days have passed by I'm snug on shore
Safe from the sea's alarms
I have a never failing store
A fifteen acre farm

Oh sweet it is to till the soil
'Neath our New England sky
And sweet when I have eased my toil
To muse on days gone by

Florida 1843

Perhaps this is related to "The Sailor Boy" in William Doerflinger's *Shanteymen and Shanty Boys*.

A fifteen acre farm sounds pretty small today, but a hundred and more years ago plenty of seamen did come home to stony little New England farms that were quite as small as that. For New England farming then was largely subsistence farming, and the money crop would very likely come from fishing. Or with only a little money invested in a share of a vessel or a mill, the retired sailor could live happily enough on his fifteen acres and perhaps raise a good sized family, too.

HEARTS OF GOLD

'Twas plowing of the raging seas
Was always my delight
While those loving landlubbers
No dangers do they know
Like we long jack hearts of gold
That plows the ocean through
Yes like we long jack hearts of gold
That plows the ocean through

They are always with the pretty girls
A-setting them fine treats
A-bursting of their pretty heads
With the work they've done in a corn field
But cutting of the grass and weeds
Is all that they can do
While we long jack hearts of gold
We plow the ocean through

OF THE SEA AND SHIPS

'Tis when the sun it does go down
They lay aside the plow
And can the work no longer stand
'Tis home that they must go
Now they got their suppers with content
And into bed they do crawl
While we long jack hearts of gold
Stand many a bitter squall

When the dark and dismal night it does come on
And the winds begin to blow
Step up step up my lively lads
Step up from down below
And every man be on our decks
Our goodly ship to guard
Step up step up my lively lads
Send down the topgallant yard

The seas they run full mountains high
Which toss us up and down
We are in the midst of dangers
For fear our ship might found
But never be down-hearted boys
We will see our girls again
In spite of all our enemies
We will plow the raging main

We'll sail to all the ports of the land
Which ever yet was known
We will bring home gold and silver boys
When we arrive at home
And we will make our courtships flourish boys
When we arrive on shore
And when our money it is all gone
We will plow the seas for more

So come all you pretty damsels
The truth you did but know

SONGS THE WHALEMEN SANG

The dangers of the raging main
From labors unto you
You would have more contempt for them
Than ever yet was known
You would hate those loving landlubbers
That always stays at home

Bengal 1832

This has no title in the *Bengal* journal so I have called it "Hearts of Gold" which seems to be the important phrase. It is possible that this is a much altered version of "The Praise of Saylors" in Stone, pp. 10-13. If so it is very old indeed. The phrase "hearts of gold" is found there, too.

There is a good version of this song, though it seems much more modern, in Harlow's *Chanteying Aboard American Ships,* pp. 219-222, called "Edgartown Whaling Song." There, too, the hardship of the sailor's life is compared with that of the "lazy landlubbers" who stay at home.

In Colcord, p. 137, it is called "The Sailor's Come All Ye." That version is taken from Eckstorm and Smyth's *Minstrelsy of Maine.*

THE TEMPEST

Cease rude Boreas blustering killer
Lend ye landsmen an ear to me
Messmates hear a brother sailor
Sing the dangers of the sea
From the bounding billows fast in motion
Where the distant whirlwinds rise
Lo the tempest and troubled ocean
There the sea contends with sky

Hark the boatswain horsely bawling
Topsail sheets and haulyards stand
Down your topsail quick be hauling
Down your staysail hand boys hand
Now it's freshening quick set braces

OF THE SEA AND SHIPS

Topsail sheets quickly let go
Luff boys luff don't make way lazy
Up your topsail nimble crew

All you on your downy beds a-sprawling
Fondly locked in beauty's arms
Fresh enjoyment venting courting
Free from all but love's alarms
Around us roars the tempest louder
Think what fears our minds enthrall
Harder yet it blows and harder
Now again the boatswain calls

Topsail yards point to the wind boys
See all clear to reef each course
Let go the foresheet do not mind boys
Though the weather shall be worse
Fore and aft now the spritsail yard
Reef the mizzen see all clear
Haul up the preventing brace sets
Down the fore yard cheer boys cheer

Around us roar dreadful thunders
Peal on peal contend and clash
On our heads fierce rain is pouring
In our eyes blue lightnings flash
One wild water all around us
All above us one black sky
Different deaths at once surround us
Hark what means that dreadful cry

The foremast's gone cries every tongue out
Twelve feet above the deck
A leak beneath the tree sprung out
Call all hands to clear the wreck
Quick the landyards cut to clear us
Come my hearties be stout and bold

Plumb the well for the leak increases
Four feet of water in the hole

While the waves o'er the ship are beating
We for wives and children mourn
Alas to them there is no retreating
Alas to them there is no return
While the leak is gaining on us
Both chain pumps chuck below
Heaven hear have mercy on us
For only that can save us now

On the lee now is the land boys
Let the guns overboard be thrown
Heave to the pumps now every hand boys
See our mizzen mast is gone
We find the leak cannot pour fast boys
We have lightened her a foot or more
Up and rig your jury foremast
She is right we are well off shore

We are all on joys now thinking
Since kind heaven has saved our lives
Come with the can boys let us be drinking
To our sweethearts and our wives
Fill it up and about the ship with it
Close to the brim your cans you fill
Of the tempest now who feels it
All our dangers we drown in wine

Galaxy 1827

For a much shorter version of this song see Duncan, p. 257, where it is called "Cease Rude Boreas," or "The Storm."

From Duncan's notes it seems that this song was popularized by George Alexander Stevens in 1754. But the melody is much older than that. Aren't the words of the song older, too? It is sung to almost the same melody that I have used for "Hearts of Gold" but without the change of time. See also *The American Musical Miscellany*, pp. 52-55.

THE CAN OF GROG

When up the shrouds the sailor goes
And ventures on the yard
The landsmen who no better know
Believe his lot is hard

Bold Jack with smiles each danger meets
Weighs anchor heaves the log
Trims all the sails belays his sheets
And drinks his can of grog

If to engage they give the word
To quarters he'll repair
Now winding in the dismal flood
Now quivering in the air

When waves 'gainst rocks do rend and roar
You'll n'er hear him repine
Though he's on Greenland's icy shore
Or burning beneath the line

When sailing orders do arrive
Bold Jack he takes his leave

My dear my sweetest Pol he cries
I pray now do not grieve

Thy Jack will take his daily can
Of grog and drink to thee
In hopes that thou will n'er forget
Thy sailor who's at sea

But should thou false or fickle prove
To Jack who loves thee dear
No more upon my native shore
Can I with joy appear

But restless as the briny main
Must heartless heave the log
Shall trim the sails and try to drown
My grief in cans of grog

Ann 1776

Grog was a mixture of rum and water which was regular issue on all British war vessels and on many British merchant vessels, as well. And withholding of the daily grog allowance was used as a punishment for minor offences and infraction of rules.

A version of this song very similar to the brig *Ann* version will be found in Kitchiner, pp. 88-89.

THE PIRATE OF THE ISLES

I command a steady band
Of pirates so bold and free
Our isle's our home my ship's my throne
And my kingdom is on the sea
Our flag is reared on our royalmast head
At all my foes I smile
No quarter I show where e'er I go
But soon my prize I take in tow

Chorus

My men are tried my bark's my pride
My men are tried my bark's my pride
I am pirate of the isles
I am pirate of the isles
I am a pirate I am a pirate
I am pirate of the isles

Bark *Stella* from a journal page
It's all about a whaling bark
That left New Bedford city

OF THE SEA AND SHIPS

I love to sail in a pleasant gale
O'er the deep unbounded sea
With a prize in view I'll bring her to
And I'll haul her under my lee
We'll give three cheers and homeward steer
While fortune on us smiles
For there's none shall cross this famous ross
Unless my flag doth strike her course

Proud Gallia's sons and Spanish dons
With zeal and ardor burned
Came o'er the sea to capture me
But back they never returned
Old England too doth me pursue
But at her threats I smile
For her ships I've taken her men I've slain
And burnt and sunk them in the main

But now in sight a ship of might
An American seventy-four
She hauls up close and hails the ross
And her broadside in she pours
But the pirate soon return the boon
And proudly doth he smile
Till a fatal ball caused him to fall
And loud his men for mercy call

Final Chorus
In the briny deep was lain to sleep
In the briny deep was lain to sleep
Was the pirate of the isles
Was the pirate of the isles
The pirate the pirate the pirate
The pirate of the isles

Cortes 1847

I had just about given up hope of finding this song in print when along comes Harlow's *Chanteying Aboard American Ships*, and there it is on pp. 172-173.

THE DEMON OF THE SEA

Come spread your sails with steady gales
And helmsman steer her right
Hoist the grim death flag the pirate cries
For a vessel heaves in sight
Run out your guns in haste bear down
From us she must not slip
Cheer cheer lads cheer we know no fear
On board the demon ship

Chorus
Then huzza for a life of war and strife
Oh the pirate's life for me

OF THE SEA AND SHIPS

My bark shall ride the foaming tide
For I am demon of the sea

Two ships of war came from afar
From Edward England's king
Go fetch he said alive or dead
The captain of the pirates bring
But his pride I shook his ships I took
And I sunk them in the wave
Six hundred and ten of proud Edward's men
Met with a watery grave

And yon ship too I mean shall sue
That ever my bark saw
For by her rig she seems to be
A British man-of-war
Give a broadside the pirate cried
Show them a pirate's fare
Fire red hot balls destroy them all
And blow them in the air

Two ships engaged in equal rage
In dreadful murderous scene
The die was cast for a ball at last
Had struck her magazine
Now one and all did stand appalled
And seemed in great despair
For the captain too and all his crew
Were blown high in the air

Final Chorus
Then no more will he ride the foaming tide
No more a dread will he be
For the pirate's dead low lays his head
In the deep and dark blue sea

Cortes 1847

I have not found any other version of this pirate song. "Edward England's King" died in 1553. The song most certainly is not that old, but might it just possibly derive from an older song that does go back to the sixteenth century?

The melody I have used here is from JFSS, vol. 2, p. 163, where the song is not "The Demon of the Sea" but "Ward the Pirate." Some of the lines in the two songs are similar, and the tune with a very slight addition does seem to fit.

THE ROVER OF THE SEA

I'm afloat I'm afloat on the fierce rolling tide
The ocean's my home and my bark is my pride
Up up with my flag let it wave o'er the sea
I'm afloat I'm afloat and the rover is free

I fear not the monarch I heed not the law
I've a compass to steer by a dagger to draw
And never as a coward or a slave will I kneel
While my guns carry shot or my belt bears a steel

Quick quick trim the sails let the sheet kiss the wind
And I'll warrant we'll soon leave the seagull behind
Up up with my flag let it wave o'er the sea
I'm afloat I'm afloat and the rover is free

The night gathers o'er us the thunder is heard
What matter our vessel skims on like a bird
What cares she for the storm-ridden main
She has braved it before and shall brave it again

The lightning gleam flashes around us may fall
They may strike they may cleave they cannot appall
With lightnings above us and darkness below
Through the wide waste of waters right onward we go

Hurrah my brave boys ye may drink ye may sleep
The storm fiend is hushed we're alone on the deep

OF THE SEA AND SHIPS

Our flag of defiance still waves o'er the sea
Hurrah boys hurrah the rover is free

Benjamin Tucker 1849

MOST BEAUTIFUL

Most beautiful most beautiful
The wind is blowing aft
And we go tilting o'er the sea
In our little warrior craft

The stars are rushing by us now
As scattered by the blast
Now all the shores are vanishing
Now all the islands past

Two things break the monotony
Of an Atlantic trip
Sometimes alas you ship a sea
Sometimes you see a ship

Trust not too much your opinion
When your vessel's underway
Let good advice bear dominion
A compass will not stray

Swan 1837

This has no title in the *Swan* journal. Most likely it is an original, but it is too good to omit.

THE SEA RAN HIGH

The wind blew hard the sea ran high
And fast the snow fell from the sky
When a little bark on our shore was cast
And all on board but one were lost

SONGS THE WHALEMEN SANG

The hardy captain oft before
Had suffered wrecks and trials sore
But still he dared the seas to brave
And leave his home for a watery grave

The other men were stout and bold
But to the wreck they could not hold
The angry waves rolled over their heads
And sunk them in the deep like lead

But one alone of all the crew
Who was both young and feeble too
Yes he though weak outlived the gale
To tell to all the fatal tale

Lotos 1833

This has no title in the *Lotos* journal. It may be a fragment of a longer song, but I cannot be sure for I have not found it in print.

THE OCEAN QUEEN

Oh list the song of the Ocean Queen
You will like the notes of her voice I ween
For the waves beat high and the dashing foam
Plays at the base of her palace home

The blue of the ocean is in her eyes
As it dashes its waves in a thousand dyes
And the crested spray of the billowy tide
Would kiss the brow of the ocean bride

She sings I love to list when the lightnings flash
Their arrowy fires in a thundering crash
Comes rattling on over the ocean afar
Like the thundering team of the god of war

I love to gaze from my pinnace high
When storm clouds roll in sulphurous sky

OF THE SEA AND SHIPS

When the fires from heaven and the blasts of war
Mingle their notes on the ocean afar

I love the storm as it rattles along
For the billowy waves is my ocean song
And of the hour with joy I ween
That I was born for an ocean queen

Sharon 1845

THE STORM WAS LOUD

The storm was loud before the blast
Our gallant bark was driven
The foaming crests the billows reared
And not one friendly star appeared
Through all the vault of heaven

Yet dauntless still the steersman stood
And gazed without a sigh
Where poised on needle light and slim
And lighted by a lantern dim
The compass meets his eye

Thence taught his darksome course to steer
And breathed no wish for day
And braved the whirlwind's headlong might
Nor once throughout that dismal night
To fear or doubt gave way

Lexington 1853

This song seems to belong to the same general class of songs as "The Pilot" and "The Beacon Light," and it was probably literary in origin. I have not found it in print.

NEPTUNE

Ho ye ho messmates we'll sing
The glories of Neptune the ocean king

SONGS THE WHALEMEN SANG

He reigns o'er the waters the wide seas his home
Ho ye ho in his kingdom we roam

He spreads a blue carpet all over the sea
O'er which our ship walks daintily
Though down at the bottom the old monarch hails
He blows the fresh wind right into our sails

Landsmen who live on the dull tame shore
Love their homes but we love ours more
Oh a ship and salt water messmates for me
There's nothing on earth like the open sea

Landsmen are green boys I have a notion
They don't know the fun that's had on the ocean
But contented they live in one spot all their lives
Like honey bees messmates they stick to their hives

What though we have storms they have earthquakes on shore
And though we have troubles they surely have more
We gather rare foods 'mong the isles of the sea
Where the tropical fruits grow there boys are we

Oh give me the ocean naught but the sea
Is a fit home messmates for hearts that are free
Oh boys ho then let us all sing
To the glory of Neptune the ocean's king

Nauticon 1848

I have not found this song in print.

On many ships crossing the line an older seaman would play the part of Neptune, holding a trident as his badge of office. Green hands were brought before him and were then initiated. They would be doused with water and suffer all sorts of other indignities, but once the ceremony was over they were considered seamen.

OF THE SEA AND SHIPS

THE DAUNTLESS SAILOR

The dauntless sailor leaves his home
Each softer joy and ease
To distant climes he loves to roam
Nor dreads the boisterous seas

His heart with hopes of victory gay
Scorns from the foe to run
In battle terrors melt away
As snow before the sun

Though all the nations of the world
Britannia's flag would lower
Her banners still shall wave unfurled
And dare their mighty power

Now see Bellona sheathes her sword
Hushed in the angry main
The cannon's roar no more is heard
Sweet peace resumes her reign

Paulina 1808

I have not been able to find this song in print. It is one of a considerable number of songs that heralded the end of the naval wars of the eighteenth century a little prematurely. For the fighting was not finally over until after Waterloo. Then the streets of every English seaport became filled with sailors looking for ships as the Admiralty decommissioned hundreds of vessels.

THE SOVEREIGN OF THE SEA

Thus thus my boys our anchor's weighed
See Briton's glorious flag displayed
Unfurl the swelling sail
Sound sound your shells ye tritons sound
Let every heart with joy resound
We sail before the gale

From a journal page
Sometimes you ship a sea
Sometimes you see a ship

OF THE SEA AND SHIPS

Chorus
See Neptune hails his watery car
Deposed by Jove's decree
Who hails a free-born British tar
The sovereign of the sea

Now now we leave the land behind
Our loving wives and sweethearts kind
Perhaps to meet no more
Great George commands it must be so
And glory calls then let us go
Nor sigh a wish for shore

A sail ahead our decks we clear
Our canvas eased the chase we near
In vain the Frenchman flies
A broadside through the clouds of smoke
Our Captain roars my hearts of oak
Fight on to gain the prize

The scuppers run with Gallic gore
The white flag struck monsieur no more
Disputes the British sway
A prize we tow her into port
And hark salutes from every fort
Hurrah my souls hurrah

Ann 1776

A LIFE ON THE OCEAN WAVE

A life on the ocean wave
A home on the rolling deep
Where the scattered waters roar
And the winds their revels keep
Like an eagle caged I pine
On this dull unchanging shore
Oh give me the flashing brine
The spray and the tempest's roar

Chorus

A life on the ocean wave
A home on the rolling deep
Where the scattered waters roll
And the winds their revels keep

Once more on the deck I stand
Of my own swift gliding craft
Set sail farewell to the land
For the wind it blows fresh abaft
And we shoot through the sparkling foam
Like an ocean bird set free
We'll find a home on the ocean wave
A home far out at sea

OF THE SEA AND SHIPS

The land fades from our view
And the clouds begin to frown
But with our bark and crew
Oh let the storm come down
And the song of our hearts shall be
While the wind and waters rave
A life on the heaving sea
A home on the bounding wave

Cortes 1847
A Hicks 1879

This is one literary song that all sailors knew and sang. And it has also been widely used as a fiddle tune. The words are by Epes Sargent of Gloucester, and if anyone could write a sea song that sailors would adopt it seems fitting that it should be a man from Cape Ann. Henry Russell composed the melody which is a good one. It was adopted as the official march of the Royal Marines by authority of the British Admiralty.

Of Sailors and Maidens Fair

THE SAILOR was a romantic figure who had crossed the raging seas to strange and distant lands. He had faced danger with bravery and even bravado, and so no sailor seems ever to have had much trouble in getting a girl, although he was usually very poor matrimonial material. And the girls who waited for the sailor were as always of two sorts; there were those who walked beside the shore with him in dewy-eyed innocence, and there were those who regarded him as a lush source of income and nothing more.

Some of these songs end on a note of "sweet content," but many end in sorrow and sadness, and perhaps that is at it should be, for sailors were realistic enough. The only songs that are lacking here are the truly bawdy ones, and make no mistake, sailors sang them. But few seem ever to have been recorded in the journals or logbooks. And those that were so recorded seem all to have been purloined with a good sharp knife.

COVENT GARDEN

It was down in Covent Garden
One day I chanced to rove
To view the finest flower
That in the garden grows
The one it was the chemink
Pink lily and the rose
Which was the finest flower
That in the garden grows
That in the garden grows

The one was lovely Nancy
Most beautiful and fair
The other was a virgin
That still the laurel wear
In hand and hand together
This lovely couple went
Resolved was the sailor
To know the maid's intent
To know the maid's intent

Altho that she did slight me
Because that I was poor
Oh no my love no not my love
I love a sailor dear

Down in Portsmouth Harbor
Our ship lies waiting there
All fitted out for sea my boys
When the wind it shall blow fair
When the wind it shall blow fair

If ever I shall return again
How happy I shall be
To have my own true love
Set dangling on my knee
And if ever I should return again
Unto my native shore
I will marry pretty Nancy
I'll go to sea no more
I'll go to sea no more

Hercules 1828

This pretty little song is from a broadside pasted in the back cover of the *Hercules* journal. There it has no title. Its proper name could be "Covent Garden" as I have it here, or "Cupid's Garden" or "Lovely Nancy." W. Chappel in *Popular Music of the Olden Times* calls it "Twas Down in the Cupid's Garden" and notes that Cupid's Garden is evidently a corruption for Cuper's Garden which was a celebrated place of amusement on the Thames.

For other versions of this song see Kidson (2), pp. 98-99, and Ford's *Vagabond Songs and Ballads of Scotland*, vol. 2, pp. 100-101.

For an earlier version of what must be the same song, see Cupid's Garden which follows this.

CUPID'S GARDEN

Come you gentlemen and ladies
I pray you to look back
And look in Sally's basket
And see there what you lack
There's fine Chiney oranges
And lemons full of juice

OF SAILORS AND MAIDENS FAIR

It will make men's mouths to water
When maidens are in their youth

Come you here my pretty maid
And sit you down by me
And a dainty Cupid's garden
I will show to thee
There is fine pinks and posies
And lilies mixed with white
I do protest I love you
My only heart's delight

Don't you say so
The fair maid said to him
You have a longing desire
To invade a maid
And when you've had your will of me
And from me you will go
To leave me here behind you
In sorrow grief and woe

Don't say you so
My charming pretty maid
For I will never leave you
So never be afraid
I will never leave you
By all the powers above
I do protest I love you
My charming turtle dove

And if unto the sea
I am forced for to go
And leave my dearest jewel
In sorrow grief and woe
There's nothing more shall trouble me
Nor run into my mind

Than for to leave my jewel
In sorrow and grief behind

I'll travel down to Portsmouth
To find my dearest dear
And if I do not find him
It will prove to me severe
He is the blooming of my eye
The comfort of all charms
And if he's buried in the sea
Let me die in his arms

Oh come says she I pray agree
Although you are my sweetheart
To take a ride along with me
To London in the (peecart)
Oh the streets of London town
They are both smooth and pretty
But now he's gone one day he's gone
And he's in London city

Leopart 1767

WILLIAM TAYLOR

William was a loyal lover
Blest with parents and good care
And his love he did discover
To a lady young and fair

OF SAILORS AND MAIDENS FAIR

Away they went for to be married
Dressed they were in rich array
But instead of being married
Pressed William was and sent to sea

She dressed herself in scarlet velvet
And after her true love did ride
With a musket on her shoulder
And a pistol by her side

And when on board this lady entered
By the name of Richard Carr
With her hand as white as lilies
All besmeared with pitch and tar

On the deck this lady clambored
Being there a many she met
There the wind did blow her waistcoat
And expose her milk-white breast

Says the captain oh good lady
What misfortune brought you here
She says 'twas for the sake of her true love
That you have pressed over here

Says the captain my good lady
What is the name of this young man
She said his name was William A. Taylor
Born and raised on the Isle of Man

If his name is William A. Taylor
Born and raised on the Isle of Man
You shall see him without a-roving
For there he walks along the strand

Herald 1817

This song is variously known as "Sweet William," "Bold William

Taylor," and there are more. It seems to have been one of the most popular songs that the whalemen sang.

It tells the familiar story of the girl who dresses like a man and follows her "sweet William" to sea. We know historically that actually a number of girls did go to sea dressed as men. In this song, what happened to sweet Sarah, and to Bold William when she found him, vary in the different versions. The *Herald* version is quite unusual in that it does not end on a note of violence.

This song has been collected frequently, and is very often found in print, but see particularly JFSS, vol. 3, pp. 214-220, and JAF, vol. 22, pp. 380-382, and vol. 28, pp. 162-163.

THE TARRY TROUSERS

As I walked out one fine May morning
The weather being fine and clear
I thought I heard a tender mother
Talking to her daughter dear

Oh daughter dear I'd have you to marry
And no longer lead a single life
Oh mother dear I would sooner tarry
For my jolly sailor bright

Oh daughter sailors are given to roaming
To foreign parts they do go
And they will leave you brokenhearted
They will prove your overthrow

OF SAILORS AND MAIDENS FAIR

No mother sailors are men of honor
For they do face the enemy
Whilst the thundering cannons do rattle
And the bullets they do fly

Mother would you have me wed with a farmer
And rob me of my heart's delight
Oh give me the lad with the tarry trousers
They shine to me like diamonds bright

Now Polly dear our anchor is weighing
And I have come to take my leave
Oh then says he my dearest jewel
Polly dear now do not grieve

Jimmy dear may I go with you
No foreign dangers do I fear
Whilst he is in the height of battle
She cries fight on my jolly tar

Come all you fair and beautious maidens
Who know a sailor is your heart's delight
I'd never have you wed with any other
For all their gold and silver bright

Nauticon 1848

THE TARRY TROUSERS
(Second Version)

Oh daughter oh daughter if ever you marry
Marry with a man that lives on shore
Pray marry with the man that you adore
A man with gold and silver in store
So don't marry with a sailor then
Not with a roving ranting wild young man

For sailors they do curse and swear
On board of man-of-war or a privateer

It's all their delight for to kill and to spoil
While landsmen are more meek and mild
So don't you marry with a sailor then
Not with a roving ranting wild young man

The landsman he has money on the land
But what the sailor gets he spends
First to an ale house then to a whore
He lives like a madman when on shore
So don't you marry with a sailor then
Not with a roving ranting wild young man

Mother sailors weather the wind and storm
To keep our country safe from harm
And when they meet with a girl that they adore
They will bring them gold and silver store
For I love a sailor as I love my life
And I mean to be a jolly sailor's wife

Oh mother you may take the landsman's part
But they'd all been sailors if they'd had heart
(Yes mother you may take the landsman's part
But they'd all been sailors if they'd had heart)
For I love a sailor as I love my life
And I mean to be a jolly sailor's wife

Oh daughter oh daughter since you are so inclined
And on a sailor have fixed your mind
Pray marry with the man that you adore
And he will give you gold and silver store
For I must allow that the most of them
Are a noble set of nice young men

Cortes 1847

In the *Nauticon* version of the song, Polly actually follows her jolly sailor to sea. But I have the feeling that this stanza is added from some other song.

OF SAILORS AND MAIDENS FAIR

See Greenleaf, pp. 69-70; Creighton (2), pp. 212-214; Sharp (1), vol. 2, p. 168; and JFSS, vol. 2, pp. 153-154 and vol. 3, pp. 313-314.

The second version, which is called "The Mother's Admonition," is evidently another version of "The Tarry Trousers." The format of this song in the *Cortes* manuscript is somewhat mixed up. Lines three and four of the fifth stanza are missing, so I have repeated line one and two of that stanza in the parentheses to make the song singable.

THE CAPTAIN CALLS ALL HANDS

The captain calls all hands and away tomorrow
Leaving our girls behind in grief and sorrow
Dry up your brimming tears and cease of weeping
How happy we shall be at our next meeting

Why will you go abroad fighting with strangers
When you can stay at home free from all dangers
For I will need you in my arms my dearest Will
So stay at home with me and your promise fulfill

Fare ye well parents father and mother
I am your daughter you have no other

And when you think on me how I am a-grieving
You see the lad that I love has proved my ruin

Down on the ground she fell like one a-dying
Lying and crying and saying there is no believing
There is no believing none not one's own brother
Excepting two can agree and love each other

Bengal 1832

This song is also called "The Bold Privateer" and "Our Captain Cried." Frank Kidson in *Traditional Tunes* says that the song must go back to the time of the Napoleonic wars at least. Personally I have a feeling that it is older than that.

See Sharp (1), vol. 2, p. 175; and also JFSS, vol. 1, p. 131, vol. 2, p. 202, and vol. 3, pp. 98-99; also, JAF, vol. 35, pp. 357-358.

A YOUNG VIRGIN

I am a young virgin just come on board
And I have as envious a maidenhead
As ever a young man took in his hand
Besides I have forty pounds in land

This young virgin as we understand
Took a trip to a foreign land
Whereas forty young lovers a-roving came
To some of their callings I long for to name

The first was a merchant that came in
He told what a traveler he had been

100

OF SAILORS AND MAIDENS FAIR

He boasted how he could handle his pen
He said he could write the best of all men

The next was a doctor that came in
He said what a traveler he had been
He said he had a lance that would open a vein
With pleasure with ease without any pain

The next was a pothecary with his pastle and pills
He said he could cure me of all of my ills
Then he took out his (glister pipe)
And I gave the rogue a mighty wipe

The next was a tailor his (body bent)
I lugged the poor rascal by both of his ears
In short I told him for to be brief
For I never intended to wed with a thief

The next was a fiddler that came in
He told what a traveler he had been
He offered for to play me a jig
I broke his fiddle and tore off his wig

I put his saddle strap out of tune
The people they surrounded the room
The tears were trickling down his face
His fiddle was broke I (potted) his case

The next was a sailor a sailor bold
With his pockets lined with gold
He (waited) not but ended the dispute
Sir here is my heart and maidenhead to boot

Herald 1817

This song may be related to the "Female Robber" See Williams, pp. 267-268.

Some of the symbolism in the song will not bear too close scrutiny,

Scrimshaw on a sperm whale's tooth
You scarce could find so fair a dame
To search this wide world over

I am afraid. Note well that in the end it is the sailor — of course — who gets the girl.

THE NOBLEMAN'S DAUGHTER

It's of a rich nobleman's daughter
So comely so handsome so fair
Her father possessed a great fortune
Of thirty-five thousand a year
He had but one only daughter
Caroline was her name I am told
One day from her drawing room window
She espied a young sailor bold

His cheeks appeared like two roses
His hair it appeared like the jet
Caroline watched his departure
Walked around and young William she met
She said I'm a nobleman's daughter
Possessed of ten thousand in gold
But I'll forsake father and mother
To wed with my young sailor bold

Said William I'd have you remember
Your parents you're bound to obey
In a sailor there is no dependence
When their true love is far far away
He said stay at home with your parents
It's do by them as you are bound

SONGS THE WHALEMEN SANG

Don't ever let anyone tempt you
To wed with a young sailor bold

She says no one ever shall tempt me
One moment to alter my mind
I'll dress and possessed with my true love
He never shall leave me behind
She dressed like a gallant young sailor
Forsook both her parents and gold
Three years and a half on the ocean
She ploughed with her young sailor bold

Three times on the ocean was shipwrecked
But always proved constant and true
Her duty she did like a sailor
Went aloft in her jacket so blue
Her father long weeps and lamenting
The tears down his cheeks they did roll
In time they arrived safe in England
Caroline and her young sailor bold

Caroline went straightway to her father's
In her jacket and trousers so blue
He received her and instantly fainted
When she appeared to his view
Shie cried my dear father forgive me
Deprive me forever of gold
But grant my request and desire
To wed with my young sailor bold

Her father admired young William
And vowed that in sweet unity
God spare his poor life till tomorrow
It's married this couple shall be
They were married on Caroline's portion
Of thirty-five thousand in gold

So now they live happy and cheerful
Caroline and her young sailor bold

Walter Scott 1840

This is another in the great family of songs in which the girl dresses as a man and goes to sea or to the wars to follow her man or to search for him. There is a quite similar version of this song in the *Cortes* manuscript (1847), and here is one stanza of it to show how close the two versions are:

It was a rich Nobleman's daughter
Both beautiful and handsome we hear
Her father was possessed of great riches
Full thirty-five thousand a year
He had one only daughter
Caroline was her name I am told
One day from her drawing room window
She admired a young sailor bold

For other versions of this song and for references see Creighton (1), pp. 66-68; and JFSS, vol. 2, pp. 181-182.

JOHN RILEY

As I walked out one evening
Down by the riverside
I heard a maid complaining
As the tears fell from her eyes
This is a cold and stormy night
These words to me did say
My love is on the raging seas
Bound to America

My love he is a sailor bold
His age is scarce eighteen
He is as nice a young man
As ever eyes have seen
My father has great riches
And Riley he is poor
And because I loved my sailor boy
They can not him endure

John Riley is the traveler's name
He lives near the town of Bray
My mother took me by the hand
And these words to me did say
If you are fond of Riley
Best let him quit this land
For your father swears he will take his life
Or else shun your company

Oh mother dear don't be severe
Where will I send my love
My heart lies in his breast
As constant as a dove
Oh daughter dear I am not severe
Here is one thousand pounds

OF SAILORS AND MAIDENS FAIR

Send Riley to America
To purchase there some ground

Now when she got the money
To Riley she did run
For to take your life this very night
My father's charged his gun
And here is one thousand pounds in gold
My mama sends to you
And sail ye off to America
And I will follow you

And when he got the money
Next day he did sail away
Before he put his foot on board
These words to him she did say
Here is a token of my true love
And I will break it in two
You will have my heart and half my ring
Until I find out you

In two or three days after
She was walking along the sea
For Riley he returned again
And took his love away
Their ship was wrecked all hands were lost
Her father grieved full sore
To find young Riley in her arms
And they drownded on the shore

It was in her bosom a note was found
And it was wrote with blood
Cruel hearted father that you are
That thought to shoot my love
So now let this be a warning
To all fair maids so gay

For to never let the lad you love
Sail to America

Cortes 1847

There are a great many versions of "John Riley" and as might be expected Riley is spelled in a great many ways, too. The much more common version of the song is related to "The Banks of Glenco," and "The Dark-Eyed Sailor" and all the other songs of that ilk in which the sailor returns after a long absence and finds his girl still faithful to him, though at first she fails to recognize her lover. The other John Riley song to which this is related, is often called "John Riley the Fisherman" or 'John Riley the Sailor."

Greenleaf, pp. 182-183 has a version much like this one called "Riley to Ameriky," and in the notes are many references. See also Kidson (2), pp. 12-13, who apologizes for "the doggerel of the verses" but stresses the beauty of the melody that he uses. The song is also in JFSS, vol. 1, pp. 256-257, and vol. 5, pp. 147-148 JAF, vol. 52, pp. 31-32, and vol. 67, pp. 127-128.

THE BRITISH MAN-OF-WAR

It was down in yonder meadow
I carelessly did stray
I overheard a damsel
And a young sailor say
He says my pretty Susan
Soon I must leave the shore
For to cross the briny ocean
On a British man-of-war

Pretty Susan fell a-weeping
Young sailor she did say
How can you be so venturesome
As to throw yourself away
For when that I am twenty-one
I shall receive my store
Oh jolly sailor do not venture
In a British man-of-war

How can you be so venturesome
As to face those brave Chinese
For they will prove as treacherous
As any Portuguese
And by some deadly dagger
You may receive a scar
So turn your inclination
From a British man-of-war

Oh Susan lovely Susan
The time will quickly pass
Let us go down to the ferry house
And take a parting glass
For my shipmates they are waiting
To row me from the shore
All for old England's glory
In a British man-of-war

Oh Susan lovely Susan
The truth to you I'll tell

The British flag is insulted
And old England knows it well
I may be crowned with laurels
Just like some jolly tar
I'll face the walls of China
In a British man-of-war

Then the sailor took his handkerchief
And cut it fair in two
Pretty Susan keep one half for me
And I'll do the same for you
The bullits may surround me
And load the cannons roar
I'll fight for fame and Susan
In a British man-of-war

A few more words together
And love let go my hand
The jovial crew they launched the boat
So merrily from the strand
The sailor waved his handkerchief
When far away from shore
Pretty Susan blessed the sailor bold
In a British man-of-war

Cortes 1847
Catalpa 1856

This seems to have been a popular and very well-known song The *Cortes* and *Catalpa* versions are much alike except that the latter does not have the seventh stanza. But in Frank Kidson's *Traditional Tunes*, pp. 102-103, there is a most interesting version, which includes some of the elements of "William Taylor," for there, Susan actually does follow her sailor to face the walls of China on a British Man-Of-War, and "She did receive a scar and she got slightly wounded."

In Galvin's *Irish Songs of Resistance* there is an adaptation of the song called "The Fenian Man O' War. See also JFSS, vol. 7, pp. 9-10.

PRETTY SALLY

There was a young sailor from Dover he came
He courted Pretty Sally Pretty Sally was her name
Her fortune it was great and her prospects were high
So that Sally on a sailor could scarce cast her eye

Oh Sally oh Sally oh Sally said he
I am afraid that your hard heart will my ruin be
Unless that your hatred should turn into love
I am afraid that your hard heart will my ruin prove

My hatred is not to you says she or any other man
But to say that I love you is more than I can
So it is cease your intentions and alter your discourse
For I never will marry you unless that I am forced

Well seven long weeks being over and passed
This beautiful damsel she took sick at last
She was entangled in love and she could not tell for why
So she sent for this sailor which she had denied

Oh Sally oh Sally oh Sally said he
I'm the young man that you would wish to see
Yes you are the doctor that can either kill or cure
Without your assistance I shall die I am sure

Oh Sally oh Sally oh Sally said he
It is don't you remember love Oh how you slighted me
Yes you slighted me my love and you treated me with scorn
And now I will reward you for what you have done

Oh what is past and gone love forget and forgive
And allow me in this world some longer to live
Oh what is past and gone I never can forgive
And I never will marry you while I have breath to breathe

She took from her fingers gold rings three
Saying keep them my love in remembrance for me
In remembrance of me my love when I am dead and gone
And perhaps you will be sorry for what you have done

Here is adieu to my father my mother and friends
Adieu unto that young man that will not make amends
Adieu unto that young man that will not pity me
One thousand times over my folly I do see

Sharon 1845

A. L. Lloyd in the *Penguin Book of English Folk Songs,* pp. 92 and 122, calls this song "The Sailor from Dover." And perhaps that is its proper title. It seems to be a rather rare song and is interesting and unusual because the situation with which it deals is the opposite of the usual one in which it is the girl who is slighted and will not forgive the man — as in "Barbary Allen" and "The Brown Girl." But here the Dover sailor will not forgive the girl.

See also, Kidson (2), pp. 20-21; JFSS, vol. 8, p. 57, where it is called "Pretty Sally." There are also a number of interesting versions of this song in *The Journal of American Folk-Lore,* where it is called variously "The Rich Irish Lady," "The Rich Lady from London" and "An Irish Young Lady from Dublin She Came." Vol. 32, vol. 39, pp. 110-11, vol. 63, pp. 259-261.)

THE BANKS OF GLENCO

As I was a-walking one evening of late
When Flora's gay mantle the fields decorate
I carelessly wandered how far I don't know
By the banks of the fountain that lies near Glenco

I spied a fair maiden she appeared like the sun
And her comely features my poor heart had won
Says I your kind affection if on me you'll bestow
We will bless that happy hour we met in Glenco

Stand off young man for your suit I disdain
I once had a sweetheart young Daniel by name
He has gone to the wars about ten years ago
And a maid I'll remain till he returns to Glenco

Perhaps that young Daniel regards not your name
And has placed his affections on some other dame
He might have got married for all that you know
And left you to wander and weep in Glenco

My Daniel my Daniel from his word never turned
For love truth and honor I found in his heart
And if he has got married for all that I know
A maid I'll remain till he returns to Glenco

He finding her constant then pulled out a glove
Which she gave him at parting as a token of love
She fell on his breast while the tears down did flow
Saying you are my Daniel returned to Glenco

Cheer up dearest Flora and break not your heart
For since we are met again no more shall we part
If the horrors of war around us should flow
In peace and contentment we'll dwell in Glenco

Catalpa 1856

THE BANKS OF GLENCO
(Second Version)

As I went a-walking one evening of late
When Flora's gay mantle the fields decorate
I carelessly wandered where I did not know
On the banks of the fountain that lay in Glenco

Like her who the prise of Mont Filo had won
There approached me a lass as bright as the sun
Ribands and tartans all around her did flow
That once was MacDonald's the pride of Glenco

I thought she was enchanted to her I drew nigh
The red rose and white lily on her cheeks seemed to vie
I asked her her name and how far she'd to go
She answered kind sir I'm bound for Glenco

I said my dear lassie your enchanting smile
And your comely sweet features does my heart beguile
If your kind affections on me you'll bestow
You'll bless the happy hour we met in Glenco

She answered me young man your suit I disdain
I once had a sweetheart MacDonald by name
He went to the wars about ten years ago
And a maid I'll remain till he returns to Glenco

114

OF SAILORS AND MAIDENS FAIR

The power of the French is hard to pull down
It's caused many a hero to die of their wounds
And with your MacDonald it may happen so
The lad you dearly love perhaps he's laid low

MacDonald from his promise will never depart
For truth love and honor are found in his heart
And if I never see him still single I'll go
And mourn my MacDonald the pride of Glenco

When I found her so constant I pulled out a glove
That she gave me at parting as a token of love
She fell in my arms while the tears they did flow
Saying you are my MacDonald returned to Glenco

Cheer up my dear Flora we have met once again
I've travelled through England through France and through Spain
Now we are together and married we'll be
And live in contentment and sweet unity

MacDonald's true valor when twice in the field
Like his noble ancestors disdaining to yield
The Spaniards and French we will soon overthrow
And in splendor return to my arms in Glenco

Romulus 1851

This song seems to have had tremendous currency. It has been collected over and over again in all parts of the English-speaking world, and it goes under many different names among which are "Donald's Return to Glencoe," "The Sailor's Return," "The Stranger and the Maiden Fair." Some versions of "John Riley" are almost identical with this song, as is also "The Banks O'Claudy." For the latter song see Ford, vol. 2, pp. 211-212.

In the great family of songs to which this belongs the hero comes home after having been away for years to find that the girl has been faithful all that time. Why she does not recognize him at once can only be explained by the fact that he was a boy when he went away, and now he returns a man, perhaps even with whiskers.

For various versions and melodies of this song see Frank Kidson's *English Peasant Songs;* JFSS, vol. 2, p. 171, and vol. 4, pp. 100-103; also Joyce, *Old Irish Music*, p. 42 and Petrie, p. 170.

THE BEGGARMAN

As young William and Mary stood by the sea side
Their last farewell, for to take
Should you never return little Mary she cried
I am sure my poor heart it would break

Do not be dismayed young William he said
As he pressed her dear hand by his side
My absence don't moan should I never return
To make little Mary my bride

Three years passed away and no news did she hear
Till she sat in her own cottage door
A beggarman came by with a patch all on his eye
Quite lame with pity he implored

Your charity said he if you'll bestow on me
Your fortune I will tell you besides
The lad that you mourn he will never return
For to make Little Mary his bride

She started and trembled at what she had heard
Saying all the money that I have I will give
If that which I ask you will tell to me true
It is where does my sweet William live

He is living said he in the greatest poverty
Although shipwrecked he has been beside
His return will be no more because he is poor
For to make Little Mary his bride

He lives Heaven knows oh the joy that we feel
Although his misfortune don't mourn
He would be welcome to me in the greatest poverty
With his blue jacket tattered and torn

The beggar hearing this off his old coat he threw
Off his old coat and his crutch too beside
With his jacket so blue and his blue trousers too
Young William stood by his Mary's side

Forgive me dear young maid young William he said
'Twas only your love for me that I tried
So to church let us away by the dawning of day
And I will make Mary my bride

Elizabeth 1845

This song belongs in the same family with "The Banks of Glenco," "The Dark-Eyed Sailor" and "The Mantle So Green." But it has the added element of intended disguise. If only the business of the ring were added one could think that "The Beggarman" was a much abbreviated version of "Hind Horn."

In Barry (2) this song is called "Little Mary, The Sailor's Bride." In Clements, pp. 106-107, it is called "William and Mary," as it is also in Barrett, pp. 58-59. See also JAF, vol. 39, pp. 114-116 and vol. 45, pp. 102-103.

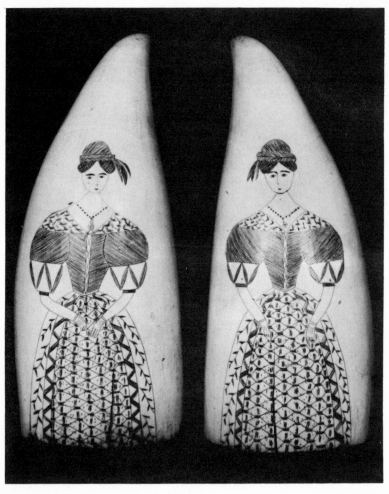

Scrimshaw on matched sperm whale's teeth
The first time I saw my love, happy was I

BRIGHT PHOEBE

'Twas in the merry month of May
When nature painted all things gay
I heard a young man sigh and say
He had lost his lovely jewel

Bright Phoebe was my true love's name
Her beauty did my heart inflame
You scarce could find so fair a dame
To search this wide world over

My love and me we did agree
That married we would truly be
When that I did return from sea
We'd seal that solemn bargain

But when I did return again
Death had my dear companion slain
Both pride and beauty of her frame
In death's cold grave lie mouldering

Oh that I never returned on shore
Or seen my native land no more
Oh cursed be that dreadful hour
That brought these dreadful tidings

I am forsaken and forlorn
I wish I never had been born

Or died where billows loud do roar
Since fortune proved so cruel

I'll go into some lonely place
Where I shall see no human face
And spend the remnant of my days
A-mourning for bright Phoebe

Come all ye landsmen list to me
And do for me some pity take
See what poor sailors do receive
From the loves they do forsake

Cortes 1847

Norman Cazden in his *Abelard Folk Song Book* says that this song is of unknown origin, but that it must have had some currency for he cites two "relatives" of it, one from Maine and one from Minnesota. There is a fragment of the song in *The Bulletin of the Folk Song Society of the Northeast,* No. 3, p. 15, where it is called "The Down-East Maid." It is the melody for that fragment that I have used here but in a slightly simplified form. And there are two versions of "Bright Phoebe" in Helen Creighton's *Maritime Folk Songs,* pp. 96-97.

THE DARK-EYED SAILOR

There was a comely young lady fair
As she walked out to take the air
She met a sailor on the way
So I paid attention to what they did say

OF SAILORS AND MAIDENS FAIR

He says pretty maiden why roam alone
The day is far spent and the night coming on
She answered him while the tears did fall
'Tis a dark-eyed sailor that is proving my downfall

'Tis three long years since he left this land
When he took a gold ring from off his hand
He broke the token here is half with me
And the other is rolling 'pon the raging sea

Cries William drive him from your mind
There is plenty of sailors left behind
Love turned aside it colder grows
Like a winter's morning when the hills are clad with snow

These words did Mary's fond heart inflame
On me she cried you shall play no game
She drew a dagger and then she cried
For my dark-eyed sailor I'll live and die

'Twas his coal black eye and his curly hair
And his flattering tongue did my heart ensnare
He was so manly no rake like you
To advise a maiden to deceive a jacket blue

But a tarry sailor I will not disdain
I will always treat them as the same
And to drink his health here's a piece of coin
But my dark-eyed sailor claims this heart of mine

Then William did the ring unfold
She seemed distracted mid joy and woe
She cried I've houses I've silver and gold
For my dark-eyed sailor so manly true and bold

In a cottage neat by the river side
It's William and Mary they do reside

So girls prove true while your lovers are away
For a cloudy morning oft brings a pleasant day

Cortes 1847

The "Dark-Eyed Sailor" seems to have been a very popular folk song indeed. Versions and variants of it will be found in many folk song collections, so many that I shall list only a few of them here. The theme of all of these "sailor returns" songs is so nearly identical that elements from one of them were very easily transferred to another. That sometimes leads to considerable confusion.

See particularly, Barry (2), pp. 42-43; Clements, pp. 104-105; and Lochlainn, pp. 10-11.

THE MANTLE SO GREEN

As I was walking one morning in May
The fields were in blossom and the meadows so gay
I espied a fair damsel she appeared like a queen
With a fine costly robe round her mantle so green

Says I pretty fair maid if you will dine with me
I will join you in wedlock and married we'll be
In fine rich attire you'll appear like a queen
With a fine costly robe round your mantle so green

OF SAILORS AND MAIDENS FAIR

Then says she young man I must you refuse
I will wed with no man so you must me excuse
Through the green woods I'll wander and I must scorn you
For the lad that I love is at famed Waterloo

Since you will not marry pray tell your love's name
For I have been in battle and might know the same
Draw near to my mantle it is there to be seen
His name is embroidered on my mantle so green

She threw back her mantle and there to behold
His name and her name too in letters of gold
It was William O'Riley it appeared in full view
He was my chief commander in Famed Waterloo

He fought for victory where the bullets did fly
On that field of honor your true love he did die
We fought for two days until the third afternoon
He received his death summons on the eighteenth of June

When he was dying these words I heard him say
If you were here Nancy content would I die
She fell into my arms with her heart full of woe
Rise up lovely Nancy your sorrows are o'er

Oh don't you remember the first time we met
It was in your father's garden I first gained your heart
It was in your father's garden where we oft time have been
When I rolled you in my arms in my mantle so green

Ocean Rover 1859

Here the returning lover is a soldier who has fought at Waterloo rather than the usual sailor, but the theme of unrecognition is the same as with this whole class of songs. And the idea of the name embroidered on the mantle is very old.

In Greenleaf, pp. 172-177 there are three interesting versions of this song, one of which is very much like that in the *Ocean Rover* journal.

OUR SHIP SHE IS LYING IN HARBOUR

Our ship she is lying in harbour
Just ready to set sail
May the heavens prove your guide my girl
Till I return again

Says the old man to his daughter
What makes you so to lament
Is there not a lad in all our town
That will prove your heart's content

There is not a lad in all our town
Nor any lord said she
For you've pressed the only lad I love
And sent him far to sea

Well if that be your inclination
The old man did reply
I hope to God he'll continue there
And on the sea will die

Then like some weeping angel
To the rocks every day went she
She is waiting there for her only love
Returning home from sea

When seven long years were passed and gone
And seventeen long tedious days

She espied a ship come rolling in
With her own true love from sea

Yonder sits a weeping angel
She is waiting there for me
Tomorrow morning to church we'll go
And married there will be

Now church being over returning home
Returning home again
Her honored father she chanced to meet
With several gentlemen

Saying if you'll forsake that man you love
And happy be once more
Ten thousand pounds I will lay down
And plenty more in store

I want not your gold or your silver
The maiden did reply
For I am married to the man I love
And I'm happy in my mind

Cortes 1847

I think the best lines in this song are, "When seven long years were passed and gone, And seventeen long tedious days." Now that is keeping a pretty close watch on the passage of time. This seems to be quite a rare song, but there is a version of it in JFSS, vol. 1, pp. 196-197.

THE SILVERY TIDE

She was a fair young creature lived down by the seaside
Of lovely form and features she was called the village pride
'Twas by a young sea captain young Mary's heart was gained
And true she proved to her Henry when he's on the raging main

'Twas in young Henry's absence a nobleman there came
He came to court young Mary but she refused the same
She says your vows are vain I love but one she cried
I love but one therefore begone for he is on the silvery tide

Then mad with desperation this nobleman did say
I'll watch you late and early I'll take your life away
I'll cause your separation this desperate villain cried
You shall sink or swim far far from him who is on the silvery tide

As he walked out one morning to take the early air
Down by the seaside he met this lovely fair
Then said this desperate villain consent to be my bride
Or I'll send your body floating down on the silvery tide

With trembling limbs she answered him my vows I ne'er can break
I dearly loved my Henry and I'll die for his sweet sake
With a handkerchief he bound her then threw her o'er the side
And a-shrieking went poor Mary down in the silvery tide

'Twas in a short time after young Henry returned from sea
Thinking to be made happy perhaps to fix the wedding day
We fear your love's been murdered her aged parents cried
Or she's caused her own destruction down in the silvery tide

Young Henry stood like one amazed at midnight then went he
To cruise the sand beach over down by the raging sea
And early the next morning young Mary's corps he spied
Floating to and fro down in the silvery tide

Seeing young Mary's corps it brought him to a stand
He then untied the handkerchief that bound her lily hand
He then untied the handkerchief that caused young Mary's doom
Young Mary so beastly murdered who scarce attained her bloom

The nobleman was taken and the gibbet was his doom
For murdering pretty Mary who had scarce attained her bloom
Young Henry so dejected he wandered till he died
And his last words were for Mary who died in the silvery tide

Cortes 1847

After this song Histed has the following notation: "Henry C. Gill on board ship *Cortes*, South Sea." Gill may have been a member of the crew of the *Cortes*, but I think it more likely that he was aboard for a gam and that Histed got the song for him.

There are versions of this song in Creighton (2), pp. 206-209; Mackenzie, p. 147; and JFSS, vol. 1, pp. 216-217.

THE UNDUTIFUL DAUGHTER

On Gosport beach I landed
That place of noted fame
I called for a bottle of brandy
For to treat my blushing dame

Her outside rigging was all silk
Her spencer was scarlet red
All day we spent in sweet content
And at night we went to bed

So early in the morning
'Twas by the break of day
I said my pretty fair maiden
What brings you down this way

I was a merchant's daughter
From London I came down
My parents turned me out of doors
Which caused me on the town

I said my pretty maiden
I'm sorry for to say
That you have come so far from home
For to throw yourself away

But no reflexions will I pass
But ever will prove true
And when from Chatham I return
Sweet girl I'll marry you

We both shook hands and parted
Tears from her eyes did flow
She seemed quite broken hearted
To think that I should go

But as a token of true love
A ring he broke in two
One half he gave to his own true love
Saying sweet lass adieu adieu

When scarce three months were over
When from Chatham he came back

Saying now my dear I can marry you
With shiners in my sack

And so to church this couple went
Their union for to tie
And may they live in sweet content
Until the day they die

Catalpa 1856

Perhaps the more common name of this song is "Gosport Beach." There is a version of it in Frank Kidson's *English Peasant Songs,* and also one in JFSS, vol. 2, pp. 262-263. Still it seems to be quite rare.

And wasn't it big-hearted of the sailor not to hold the pretty maiden's past against her, and to pass no reflections?

THE SHIP CARPENTER

In Gosport of late there a damsel did dwell
For wit and for beauty she did many excel
A young man he courted her to be his dear
And he by his trade was a ship carpenter

He says my dear Molly if you will agree
And then will consent for to marry me
Your love it will ease me of sorrow and care
If you will but marry a ship carpenter

With blushes more charming than roses in June
She said Sweet William for to wed I am too young
Young men they are fickle and 'tis very plain
If a maid she is kind they will quickly disdain

The most beautifulest woman that ever was born
When a man has in (yord) her her beauty he'll scorn
Oh my dear Molly what makes you say so
Thy beauty is the heaven to which I will go

It's now in yon channel and church for to steer
There I will cast anchor and stay with my dear
I ne'er shall be (clide) with the charms of my love
Thy love is as true as the true turtle dove

All I do crave is to marry my dear
And arter we are married no dangers we will fear
The life of a virgin Sweet William I prize
For marriage brings trouble and sorrow likewise

But all was in vain though his suit she did deny
Yet he did oppress her for love to comply
By his cunning her heart he did betray
And with too lude lust he led her astray

This past on a while and at length you will hear
The king wanted sailors and to sea he must steer
This grieved the fair damsel almost to the heart
To think of her true love so soon she must part

She says my dear William as you go to sea
Remember the love that you made unto me
With kindest expressions he to her did say
I will marry you Molly ere I go away

If that on tomorrow to me you will come
Then we will be married and our love carried on

With kindest embraces they parted that night
She went to meet him next morning by light

He says my dear charmer you must go with me
Before we are married a friend for to see
He led her through groves and valleys so deep
That this fare damsel began for to weep

She says my dear William you lead me astray
On purpose my innocent life to betray
And that is a trueness and none can you save
For all of this night I've been digging your grave

A grave there she see and said standing by
Oh must this grave be a bride's bed to me

Vaughn 1767

This song is also sometimes called "The Gosport Tragedy." There seems to be a refrain, but I cannot tell by the very confused format in the journal just how it goes. Perhaps it consists only of the last line of each stanza repeated.

My wife used to know a local version of this song. Unfortunately it is lost because in our ignorance we combined it with a version of "Lone Green Valley" that I had learned in Florida. Now we can no longer separate the two songs. In her version the man was Sweet William and the girl was Mary.

See Creighton (2), pp. 114-120 for four interesting versions. Also, Sharp (1), vol. 1, pp. 317-327; Mackenzie, pp. 96-97; and JFSS, vol. 1, pp. 172-173; and JAF, vol. 20, pp. 259-264.

THE PRIDE OF KILDARE

When first from sea I landed I had a roving mind
Undaunted I did ramble my true love for to find
When I met pretty Susan her cheeks were like a rose
And her bosom was fairer than the lily that grows

Her dark eyes did sparkle like the bright stars at night
And the robes she was wearing was costly and white
Her bare neck was shaded with her long raven hair
And they call her pretty Susan the pride of Kildare

Long time I courted her till I wasted my store
Then her love turned to hatred because I was poor
Says she I have another whose fortune I will share
So begone from pretty Susan the pride of Kildare

It was early the next morning as along I did stray
I met pretty Susan with a young lord so gay

And as I passed by them with my mind full of care
I sighed for pretty Susan the pride of Kildare

Once more upon the ocean I am resolved to go
I am bound to the eastward with my mind full of woe
There I will behold ladies in jewels so rare
But none like pretty Susan the pride of Kildare

Sometimes I am jovial sometimes I am sad
When I think she is courted by some other lad
But as we are at a distance no more will I despair
So my blessings on my Susan the pride of Kildare

Cortes 1847
Ocean Rover 1859

The Cortes and Ocean Rover versions of this song are very similar, but there is enough variation to show that one at least was not copied from a broadside. The theme of "The Pride of Kildare" is much the same as that of "The Lily of the West" which follows, except that here the sailor leaves the girl to his rival without any violence.

Like "Black-Eyed Susan" this seems to be a folk song that was adopted by polite society, and so has become quite standardized. As a result it will be found in many popular song collections but not in many collections of folk song. However, for one version of it see JFSS, vol. 6, pp. 11-12.

THE LILY OF THE WEST

When first I came to England
Some pleasure for to find
I spied a pretty fair maid
Most pleasing to my mind
Her rosy cheeks and rolling eyes
Like arrows pierced my breast
And they called her lovely Flora
The lily of the west

Her golden hair in ringlets hung
Her dress was spangled o'er
She had rings on every finger
Brought from some foreign shore
She would entice both kings and princes
So costly was she dressed
She's the fair exile of Kennis
The lily of the west

I courted her a while
I thought her love to gain
But soon she turned her back on me
Which caused me much pain
She has robbed me of my liberty
She has robbed me of my rest
And I roam alone for Flora
The lily of the west

One day as I was walking
All in a shady grove

OF SAILORS AND MAIDENS FAIR

I spied a lord of high degree
Conversing with my love
She sang a song melodiously
While I was sore oppressed
He said adieu to Flora
The lily of the west

I stepped up to my rival
With a dagger in my hand
I snatched him from my false love
And boldly bade him stand
Being mad with desperation
I swore I'd pierce his breast
For I was betrayed by Flora
The lily of the west

It's now I stand my trial
Most lonely for to plea
There was a flaw in the indictment found
Which quickly set me free
For a beauty high I did adore
I'd kill who would molest
And I'll roam alone for Flora
The lily of the west

It's now I've got my liberty
A-roaming I will go
I'll travel through old England
I'll roam old Scotland through
For a beauty high I did adore
And she still disturbs my rest
And I'll roam alone for Flora
The lily of the west

Thomas Perkins 1844

The full title of this song in the *Thomas Perkins* journal is "Lady Flora the Lily of the West." The girl's name changes in different versions of the song as one would expect. In Sharp (1), vol. 2, p. 199, it is Mary. In Lochlainn, pp. 184-185, it is Molly-o. One would sort of like to know what the flaw in the indictment was that so quickly set the hero free.

See also, Creighton (1), pp. 84-86; and Kincaid (1), p. 46. Also JAF, vol. 36, p. 369 where the song is mentioned, with notes.

THE MAID ON THE SHORE

There was a young lady disappointed in love
And she was sunk deep in despair-o
Nothing could she find to relieve her mind
But to roam all alone on the shore-o

There was a sea captain that plowed the salt sea
The sea it ran calm and clear-o
I shall die I shall die this sea captain cried
If I can't get that young lady fair-o

Oh what shall I give you my bonny brave boys
For all for to fetch her on board-o
Some magical art has got into her heart
Makes her roam all alone on the shore-o

Our captain got pearls and diamonds and rings
And he has got costly attire-o

If you'll get her on board just as quick as you can
I will give you a run around the shore-o

With many a persuasion she entered on board
The captain he used her so fair-o
He invited her down to his cabin below
So it's farewell to sorrow and care-o

I will sing you a song this fair maiden did cry
So the captain he sat her a chair-o
She sung him a song so loud and complete
That the sea boys they all fell asleep-o

She took all his pearls and diamond and rings
And likewise his costly attire-o
The captain's broad sword she used for an oar
And paddled her boat to the shore-o

Now were my men sober or were my men drunk
Or were they sunk deep in despair-o
For to let her get away with her beauty so gay
For to rove all alone on the shore-o

Ocean Rover 1859

This song is also known as "The Sea Captain" Cazden in his *Abelard Folk Song Book*, pp. 120-121 of part 2, gives a long and scholarly discussion of the background and antecedents of this song with special attention to possible supernatural elements. See also Greenleaf, pp. 63-64, and Barry (2), pp. 40-41.

LOVELY CAROLINE

Come all young men and maidens attend unto my rhyme
It was of a virgin seated all in her prime
She bids the blushing roses admiring all around
For it was lovely young Caroline of Edingborough town

Young Henry was a Highland lad a-courting to her came
And when her parents heard of it they did not like the same
Young Henry was offended and unto her did say
Arise my lovely Caroline and with me run away

We will go to London love and there we will marry with speed
And there young lovely Caroline shall have happiness indeed
She being enticed by Henry put on her other gown
And away went young Caroline of Edingborough town

Over lofty hills and mountains together they did roam
Until they came to London far from her happy home
She said my dearest Henry pray never on me frown
Or you'll break the heart of Caroline of Edingborough town

They had not been in London not past half a year
Before hard hearted Henry proved unto her severe
It's I won't go to see your friends for they did on me frown
So beg your way without delay to Edingborough town

A gallant fleet is fitting out with speed a-dropping down
And I will go and I will fight for king and for crown
The jolly tars may fill the skies and in the waters drown
Before I will return again to Edingborough town

OF SAILORS AND MAIDENS FAIR

Yet many a day it passed away in sorrow and despair
Her cheeks were like the roses now grown to the lilies fair
Oh where now is my Henry oft times she did frown
'Tis sad the day I ran away from Edingborough town

Unto the woods now this fair maid now this fair maid did go
To eat such food as on the bushes in the woods did grow
Some strangers they did pity her and some did on her frown
And some did say why did you stray from Edingborough town

'Twas there under some lofty oaks this maid sat down to cry
A-watching of the gallant fleet as they were passing by
She gave three shrieks for Henry and plunged her body down
And away went poor Caroline of Edingborough town

Likewise she left her bonnet there upon the shore
And in a note a lock of hair and these words I am no more
I am in the deep and fast asleep and the fishes watching round
And away floated young Caroline of Edingborough town

Elizabeth 1845

Eloise Linscott in *Folk Songs of Old New England* has a version of this song, (pp. 183-185) called *Caroline of Edinboro Town* which is probably the proper title. She states that no printed versions of this song can be found earlier than the early part of the nineteenth century. But she thinks that the song may well be older than that.

In the *Sharon* journal (1845) there is a twelve stanza version of the song quite similar to both Miss Linscott's and the one in the *Elizabeth* journal. Here are a few stanzas from the *Sharon* version to show changes:

Come all young men and maidens attend unto my rhyme
It's of a young damsel who was scarcely in her prime
She beat the blushing roses and admired all around
It is lovely young Caroline of Edingburg town

The fleet is fitting out to Spithead dropping down
And I will join the fleet to fight for king and crown

From a journal page
And when I'm gone love think of this
When will we meet again?

The gallant tars may feel the scars or in the waters drown
Yet never will return again to Edingburg town

She cried where is my Henry and after did swown
Crying sad's the day I ran away from Edingburg town
Oppressed with grief without relief the damsel she did go
Into the woods to eat such fruit as on the bushes grow

Come all you tender parents never try to part true love
You're sure to see to some degree the ruin it will prove
Likewise young men and maidens n'er on your lovers frown
Think on the fate of Caroline of Edingburg town

This seems to have been a very popular song in its time, but every-
where the spelling, and probably the pronunciation of Edinburgh, too,
changed according to the taste and education of the singer. Also the song
seems to have been most popular on this side of the Atlantic.

For other versions see Mackenzie, pp. 94-95; Sharp (1), vol. 1,
p. 404; and Creighton (3), pp. 99-100; JAF, vol. 52, pp. 14-15.

THE TURKISH LADY

You young and bold I pray draw near
And a pretty story you shall hear
It was of a Turkish young lady brave
Who fell in love with an English slave

He was a merchant's young son of late
As he was sailing up the Straight

SONGS THE WHALEMEN SANG

It was by kind fortune he came to be
A slave unto some rich lady

She bound him in chains and fetters strong
She whipped and scourged him all along
It's no tongue can tell that I am very sure
The hardships that we poor slaves endure

She does dress herself all in rich array
And walked forth to view her flowers one day
Hearing a moan that the young man made
It's to him she goes and there she said

O what country man what country I pray are you
An English man madam it is true
Ah I wish you was some Turk said she
I'd free you from your slavery

I'd free you from your slavery work
If that you would but turn Turk
And I would yield myself your lawful wife
For I do love you as my life

O madam O madam then said he
That I'm sure can never be
I'd sooner be burnt unto a stake
Before my God I will forsake

She straight unto her chamber went
And spent that night in discontent
Young Cubed with his piercing dart
Soon gained this fair young lady's heart

Now to all her friends she bid adieu
For she loves an English man 'tis true
O now she is landed on the English shore
With silver and gold in great store

OF SAILORS AND MAIDENS FAIR

And now she is turned to an English dame
And married unto one of her slaves

Two Brothers 1768

This song has no title in the *Two Brothers* journal, however, its common name is "The Turkish Lady." Tristram P. Coffin in his commentary in Helen Hartness Flanders *Ancient Ballads Traditionally Sung in New England* (vol. 2, p. 10) calls it a near relative of "Lord Bateman" which is the name usually given to "Young Beichan" (Child 53) on this side of the Atlantic. The melody that I have used here is from Kidson (1), p. 33, where the song is also called "Lord Bateman."

Helen Creighton in her *Songs and Ballads from Nova Scotia*, pp. 26-28, has a nice version of the song called "The Turkish Rover." In her note for the song she says that it is probably an inferior rewriting of "Young Beichan." All scholars seem to be agreed that "The Turkish Lady" and "Young Beichan" are related, the only question being how closely.

For other versions of this song see Cecil Sharp's *English Folk Songs* (Centenary Edition), pp. 17-19; Sharp (1), vol. 1, pp. 77-88; and Williams, pp. 147-149. Also JFSS, vol. 1, p. 113, and JAF, vol 20, pp. 251-252, and vol. 28, pp. 149-151.

Of Yankee Manufacture

T HE LIST of songs of Yankee manufacture is pretty short, and even of those included here, three or four probably had their roots in the British Isles. Here there are no lumber camp songs as such, nor cowboy songs, nor songs of the wicked city. Of course most of the whaling songs were "American built" as were many of the parlor songs and music hall songs. But perhaps most of the songs that the whalemen sang came from the British Isles.

THE TIMES

The present times are deuced bad
And still worse they are getting
Father and mother is very sad
And with the children fighting

Chorus

Yankee doodle devil's to pay
Ships and produce rotting
Can't get rich by night or day
Mischief sure is plotting

I guess as how I've found it out
And father thinks likewise sir
'Tis but to please that treacherous lout
That Boni that we die sir

You've elbow room to frighten him
But Congress don't that way view
They do but black and whiten him
Act soon or long this day rue

Experiments unsuccessful still
Where sported long our stout ship
'Tis time for us to gain a port
Or heave in stays and 'bout ship

No more French palaver then
When by all hands and watched for

....

....

Had Captain Washington been found
Upon the quarterdeck sir

D'ye think he'd run the shipping round
At Boniparte's beck sir

Final Chorus
Yankee doodle no not he
Had such a pirate chased him
He'd quick put up the helm alee
And quickly trimmed and laced him

Polly 1804

Innumerable songs and verses have been set to the good old tune
of "Yankee Doodle," as has this. The "Times," here, were when the
little Yankee vessels were trying to trade with both France and England,
who were at war with each other, and get rich in the process. But Congress and the President often had other ideas.

I am sorry to have had to leave two whole lines out of the fifth
stanza, but I simply could not decipher them. Anyone who wants to
sing the stanza will have to go to the source and do better than I or make
up his own lines.

THE SONS OF LIBERTY

OF YANKEE MANUFACTURE

Come and listen sons of freedom
Hear what tidings comes each day
Concerning of Brittan's children
The troubles in America
Britons against Brittany is fighting
Blood and slaughter quickly reign
France and Spain is this delighting
Britons a-mourning for pain

It was at Lexington blood run pouring
Crimson streams on Bunker Hill
With crashing arms and cannon roaring
Where our Brittons their blood did spill
Beside (Salem) a sea engagement
Salem's (shame) was ever the day
Lament lament each worthy Briton
The life that is lost in America

New York by this time was in our hands
If that we believe the news
But how shocking is our fighting
The more we gain the more we lose
The more we best 'twill the longer be
Before our trade will shine again
Kind heaven be our moderator
Let not our blood be spilt in vain,

Oh heavens once more friendly join us
Before today we spill the blood
With hand and heart ever entwine us
It is where . . . . ? . . . for good
Let not our foes rejoice in telling
How we do each other slay
Stop shedding blood we do beseech thee
Of Britons in America

This is the wish of poor Britons
Likewise widows maids and wives

That's left to mourn sons husbands sweethearts
Was as dear to those as life
But give us peace and sweet contentment
Drive all trouble far away
That commerce in trade may once more flourish
With Britons in America

> The next affair where
> We made our landing good
> Some thousands there we were informed
> Was basely killed in cold blood
> But let us hope for the future
> They quarter to each other show
> And as we was once friends together
> Be watchful of the fatal blow

Dolphin 1790

The title for this song in the *Dolphin* journal is "A Song of Late." It is one of the very few traditional songs that show the loyalist point of view so very clearly. There were many loyalists on Nantucket during the Revolution. For many there were Quakers, and they, naturally, would be opposed to the war, but also the war had brought the whaling industry to a complete standstill.

There are two pretty badly garbled versions of this song in Sharp (1), vol. 2, pp. 224-225, but from the *Dolphin* version it looks as though the song had been garbled almost from the beginning. There is also a version of it in Kidson (1), pp. 61-63 called "The Gown of Green." And it is Kidson's melody that I have used here. Kidson says that the verses produced at the time of the last (sic) American war were grafted onto the old folk song

See also JFSS, vol. 2, pp. 90-91.

THE LASS OF MOWEE

148

As I was a-roving for pleasure one day
For sweet recreations and sore cast away
I sat in a tavern and by me a glass
There happened to come in a young Indian lass

She stepped up to me and took hold of my hand
Saying you look like a stranger away from your land
But if you will follow you are welcome to come
And I live by myself in a snug little home

Just as the sun set behind the blue sea
I wandered alone with my little Mowee
Together we rambled together we rove
Until we came to her house in a coconut grove

With fondest expressions she said unto me
If you will consent to live along with me
And never shall go roving upon the salt sea
The language I'll learn you is of the Isle of Mowee

To which I replied that never can be
For I have a sweetheart in my own country
And I never will forsake her in her poverty
She has a heart that is as true as the lass of Mowee

149

Early next morning by the dawn of the day
I grieved her to the heart when these words I did say
I am going to leave you so farewell my dear
My ship has weighed anchor and for home we will steer

The last time I saw her she was down on the strand
As my boat passed by her she waved her hand
Saying when you get home to the girl that you love
Remember the maid in the coconut grove

Now I am safe landed on my own native shore
My friends and relations gather round me once more
Not one that comes round me not one do I see
That can be compared with the lass of Mowee

This young Indian was handsome she was modest and kind
She acted her part to the heavens devine
For when I was a stranger she took me to her home
So I'll think on the Mowee as I wander alone

Cortes 1847

There is a good version of this song in the *A Hicks* journal. It is quite similar to this, but has one less stanza. There the last line of the fourth stanza reads "I will teach you the language of the Isle of Mohee." Also there is one lone stanza of the song in the *Euphrasia* journal that goes:

Now this young Indian was both honest and kind
She has acted to me as heaven designed
For I was a stranger and she took me to her home
And I'll think on that Mohe as I wander alone

Thus the name of the island and of the fair Indian maid is spelled variously Mowee, Mohee and Mohe. I point that out because the island was undoubtedly Maui in the Hawaiian group. And this song, just as undoubtedly, at least in its present form, originated with the whalemen. Barry in the *Bulletin of the Folk Song Society of the Northeast*, No. 6, pp. 15-18 does not seem to agree with that statement.

He says, "As the Kanakas were not Indians we conclude that Maui is adventitious, that the ballad originally dealt with a romance of an Indian and a pioneer."

Perhaps so, but this is not the original song, "The Miami Lass," if that is indeed the original, any more than "The Boston Burglar" is "Botany Bay." This is "The Pretty Maid of Mohee" or "The Little Mohee" and it is whalemen's work. For to a whaleman a Polynesian and an Indian would be pretty much the same thing. There were few ethnologists among them.

For other versions of this song besides Barry, see Mackenzie, pp. 154-156; Colcord, pp. 199-200; Kincaid (1), p. 38; Kidson (1), pp. 109-111; JFSS, vol. 2, p. 262; Clements, pp. 108-109; and Fuson, p. 84; JAF, vol. 39, pp. 132-134; vol. 45, pp. 96-99.

THE HEATHEN DEAR

'Tis night where strays my heathen dear
Why does she from me roam
For well she knows my heart is drear
When she is from the ship my home
But soft what music greets mine ear
What strain comes o'er the dell
Oh sweet to me the night winds bear
That sound her savage yell

Oh send the boat the queen of night
Walks she in the sparkling mountain rills
And spreads her fairest robes of light
To guide her through the dewey hills
She comes she comes her voice I hear
Her pretty form I see
And soon they'll bear my heathen dear
In joy again to me

Motuky nunay nunay kemi
Wita wita hushan hushan
The wihi nua nuka hera
The Marques Islands and

Mr. Spannen with 4 C of tons
In one whaleboat returning
To the ship to mi mi mou mou
Tetua nua nua nue quen
Tippin nua nua appo kemi
Appo ship cushany

Bengal 1832

Almost surely this delightful little original was sung, but to just what tune I cannot say. It seems somehow related to "The Moon Is Brightly Beaming Love" which follows, and perhaps also to "Farewell to Maimuna." For the latter song see Whall, pp. 3-4.

The Kanaka in the third stanza is evidently pretty badly garbled, for Samuel H. Elbert, professor of Pacific languages at the University of Hawaii tells me that he can make almost no sense of it. He thinks that the third line may mean "The Nukahivan woman." Nukahiva is the largest northern island of the Marquesas group. And "Tetua" he says is the most common feminine name in those islands.

It was only a few years after the voyage of the *Bengal* that Melville wrote his famous classic *Typee* about the Marquesas. Yankee whalemen did love those girls of the Pacific islands.

THE MOON IS BRIGHTLY BEAMING LOVE

The moon is brightly beaming love
Far o'er the deep blue sea
A trusty crew are waiting near
For thee dear girl for thee
Then leave thy downy couch my love
And with thy sailor flee
My gallant bark shall bear thee safe
Far o'er the deep blue sea
Far o'er the deep the deep blue sea

The storm birds sleep upon the rocks
No angry surges roar
No sounds disturb the tranquil deep
Not even the dripping oar

OF YANKEE MANUFACTURE

No eye beholds thee coming now
Come dearest fly with me
And cheer a daring sailor home
Far o'er the deep blue sea
Far o'er the deep the deep blue sea

She comes she comes with trembling steps
Oh happy shall we be
When landed safe on other shores
From every danger free
Then speed thee on my gallant bark
Our trust is all in thee
Safe bear us to our peaceful home
Far o'er the deep blue sea
Far o'er the deep the deep blue sea

Cortes 1847

SHEARING DAY

Now this is shearing day alack
And here we are around Cape Horn
And we shall surely miss of this
As sure as we are born

Now half the town and all bull lovers
Drive up their sheep together
And in the sheepfold shorn are they
Each ram and ewe and wether

Now some of the lads about the town
Do make a vigorous onset

SONGS THE WHALEMEN SANG

And Jehu like away they drive
With girls to Siasconsett

While others to the shear-pen go
And round the tents do caper
And dance and cut all kinds of quams
Before a cat gut scraper

The tents are filled with cakes and wine
And liquors in galore
Of beef and pork and pigs and fowl
They have abundant store

All nicely cooked and all served up
As rich as milk and honey
Where you can sit and eat your fill
As long as you have money

But as the sun keeps going down
The steam begins to rise
And 'tis quite common there to see
Red noses and bunged eyes

Now sable night her curtain spreads
And rather cool the weather
And beaux and girls begin to think
Of jogging home together

And he whose purse is fairly out
On foot to town must tag on
But he can ride who's flush with cash
In coach or cart or wagon

Now all the Siasconsett folks
Drive into town like thunder
And rattling o'er the pavement they
Make gawkies stare and wonder

Some with a broken chaise tied up
Some killed their horse a-racing

The Morn of Life.

Scrimshaw on a sperm whale's tooth
I espied a fair damsel
She appeared like a queen

But all such things on shearing day
There sure is no disgrace in

Now sing long celebrate the day
We'll dine and dance and spree it
And next year when that day comes round
May I be there to see it

Maria 1832

Shearing day on Nantucket seems to have been something of both community endeavor and carnival. And, too, it is interesting to see that in the early eighteen thirties, the time of the island's greatest prosperity, half the town still kept sheep. That must mean that most of the wool that was shorn on shearing day was still washed, carded, spun, and woven at home.

Charles Murphy wrote this when he was the first mate of the *Maria*, but if he had been home on Nantucket he would have been there around the tents capering with the best of them and scraping the cat-gut, too, for he was a great fiddle player. He was homesick.

In William Alexander Barrett's *English Folk Songs* there is a song called "Sheep Shearing Day," and the very simple melody fits Murphy's song perfectly. The melody I have used here is from JFSS, vol. 6, p. 15.

SARAH MARIAH CORNELL

Kind Christians all I pray attend
These few lines that I have penned
While I relate the murdered fate
That did await poor Cornell's end

Miss Sarah Cornell was her name
(Whom) base deceit had brought to shame
Your hearts in sympathy must bleed
When shepherds murder lambs indeed

To reverend Mr. Avery sure
A teacher of the gospel pure

OF YANKEE MANUFACTURE

Stands charged with murder to the test
Seduction too in part confessed

First in (griest) he was set at large
From circumstances there's further charge
Soon after that the deed was done
He ran away the law to shun

But beloved for blood loud doth cry
All murderers too must surely die
Three hundred dollars of reward
(Naming) this Avery to the charge

He soon was taken and with speed
Must answer to the fatal deed
Now in Rhode Island bound is he
'Tis May to await his destiny

Methought I heard her spirit say
Remember Cornell's end I pray
And let no one reflection make
Upon my friends . . . ? . . . sake

Let woman's weakness plead my cause
When cruel men break nature's laws
Oh a man by a man is much deceived
What tongue would not my weakness plead

Knew you but half the artful way
My base destroyer led me astray
The best may slip the cautious fall
He's more than man ne'er erred at all

Ye maidens all both old and young
Trust not to man's false flattering tongue
To know a man pray know his life
How few there are deserve a wife

Though doomed I am to awful end
I crave some prayers of every friend
That my poor spirit may be behest
And with my God in heaven rest

Yet I conclude my mournful song
These lines I say remember long
Adieu my friends pray don't repine
Examples yours experience mine

Sharon 1845

I wish I might have had time to search out the details of this tragedy. It probably occurred not too long before 1845 and there are surely
records of it in the papers of the time. The fact that the villain of the
piece was a minister of the gospel must have added spice to the telling
and singing of the story. Cornell is a fairly common family name in the
New Bedford area.

JOHN BROWN

Old Aunt Becky won't you set 'em up again
Old Aunt Becky won't you set 'em up again
Old Aunt Becky won't you set 'em up again
As we go marching home

Chorus

Glory glory hallelujah glory glory hallelujah
Glory glory hallelujah as we go marching home

John Brown's body lies a-mouldering in the grave
John Brown's body lies a-mouldering in the grave
John Brown's body lies a-mouldering in the grave
As we go marching home

He's gone to be a soldier in the army of the Lord
He's gone to be a soldier in the army of the Lord
He's gone to be a soldier in the army of the Lord
As we go marching home

John Brown's napsack it is strapped upon his back
John Brown's napsack it is strapped upon his back
John Brown's napsack it is strapped upon his back
As we go marching home

John Brown's pet lamb will meet him on the way
John Brown's pet lamb will meet him on the way
John Brown's pet lamb will meet him on the way
As we go marching home

John Brown's napsack it is number eighty-four
John Brown's napsack it is number eighty-four
John Brown's napsack it is number eighty-four
As we go marching home

Going down south by way of Baltimore
Going down south by way of Baltimore
Going down south by way of Baltimore
As we go marching home

We'll hang Jeff Davis on a sour apple tree
We'll hang Jeff Davis on a sour apple tree
We'll hang Jeff Davis on a sour apple tree
As we go marching home

Now the whiskey bottle it lies empty on the shelf
Now the whiskey bottle it lies empty on the shelf
Now the whiskey bottle it lies empty on the shelf
As we go marching home

Midas 1861

This version of "John Brown" seems very close to the event, or it could have been put in the journal at the end of the voyage or even after the voyage had ended.

The first stanza here is not in the journal, but from my childhood. I have included it because it ties in so nicely with the last stanza. For a lot more stanzas for "John Brown" see John and Alan Lomax, *American Ballads and Folk Songs*, pp. 528-529.

THE BANKS OF THE SCHUYLKILL

On the banks of the Schuylkill so pleasant and gay
There blest with my true love I spent the short day
Where the sun did peep through the mulberry tree
And the stream formed a mirror for my true love and me

On a spot of green clover we set ourselves down
Not envying the greatest of monarchs that's crowned
My name in the sand with his finger he drew
And he swore by the stream he would ever be true

Then I beheld the gay pride of my fair
I gazed on his face while he played with my hair
He need not have told me his love with a sigh
For the Schuylkill binds my dear fellow to me

My lips were solicited my hand gently pressed
On the banks of the Schuylkill where Jesse was blest

OF YANKEE MANUFACTURE

Whenever we leave this enchanting retreat
With blushes she says when next shall we meet

Next Sunday he says if the weather prove clear
On the banks of the Schuylkill I'll meet you my dear
Now all these innocent pleasures are o'er
And the mourning river can please me no more

Now the banks of the Schuylkill have lost all their charms
For the soldiers have taken my love from my arms
But should I ever clasp him again to my heart
No more should my true love and I ever part

Oft times he told me the stories of love
He would sing me a song my affections to move
When the wars shall no more take my true love away
The banks of the Schuylkill shall ever be gay

Fortune 1840
Chile 1843

I have not seen this song in print, but the fact that it is found in two journals must indicate that it had some currency. Both versions are quite similar.

It was Rudolphus Dexter of West Tisbury on Martha's Vineyard who kept the *Chile* journal. Under the date Saturday, November 19th he says:

Oh could I but my dear little wife and children see
How pleasant the banks of the Schuylkill would be.

THE BANKS OF CHAMPLAIN

'Twas autumn and round me the leaves were descending
And lonely the woodpecker pecked on the tree
Whilest thousands their freedom and rights were defending
The din of their arms sounded dismal to me
For Sandy my love was engaged in the action
Without him I valued the world not a fraction

161

SONGS THE WHALEMEN SANG

His life would have ended my life in distraction
As lonely I strayed on the banks of Champlain

When turning to list to the cannons' loud thunder
My elbow I leaned on a rock near the shore
The sounds nearly parted my heart strings asunder
I thought I should see my dear shepherd no more
But soon an express all my sorrows suspended
My thanks to the fathers of mercies ascended
My shepherd was safe and my country defended
By freedom's brave sons on the banks of Champlain

I wiped from my eye the big tear that had started
And hastened the news to my parents to bear
Who sighed for the loss of relations departed
And wept at the tidings that banished their care
The cannons now ceased and the drums were still beating
The foes of our country far north were retreating
And neighboring damsels each other were greeting
With songs of delight on the banks of Champlain

One squadron triumphant our army victorious
With laurels unfaded our Spartans returned
My eyes never dwelt on a scene half so glorious
My heart with such rapture before never burnt
But Sandy my darling that moment appearing
His presence to every countenance cheering
Was rendered to me more doubly endearing
By feats he performed on the banks of Champlain

Nautilus 1838

There are a number of songs dealing with the battle of Lake Champlain, but I have not yet found another version of this one. Beside the title there is this notation: "By Mrs. M." And after the last stanza this: "1839 This is the end of the song."

162

OF YANKEE MANUFACTURE

A NEW LIBERTY SONG

Awake awake ye Americans
Put cheerful courage on
If tyrants then shall oppress you
Arise and say begone

Oh let no papis bear trickery
Nor tyrants ever reign
Treat such infringments of our rights
With insolent disdain

Yet we will loyal subjects be
To any loyal king
And in defense of such a prince
Spend every precious thing

But when our prince a tyrant grows
And Parliament grows worse
New England's blood shall never bear
The ignorminious curse

Then let Lord North and Hutchinson
And Barnard do their work
Their hated names through every age
Forever shall be curst

But mortal tongue cannot express
The praise that shall abound
Upon the head of everyone
Who proves New England's friend

Though navies do around us lie
And troops invade our land
Yet we'll defend our liberty
As long as we can stand

Since fighting is our last redress
We'll bravely let them know

SONGS THE WHALEMEN SANG

That we will fight with all our might
Before our rights shall go

All for the sake of liberty
Our fathers first came here
And hunger here went with the cold
And hardships most severe

Then let no mighty tyrant think
We're such a wretched brood
As to give up that liberty
Our fathers bought with blood

We gladly will consent to peace
At reasonable terms
Our liberty once well secured
We will lay down our arms

But never will resign those rights
Our fathers purchased so
Whilst any of their noble blood
Within our veins does (grow)

Domestic enemies we have
Almost in every town
Whose names to unborn ages shall
Be always handed down

With infamy dishonor's yoke
Shall link them in disgrace
Amongst the sons of liberty
Till time itself shall cease

Unite unite Americans
With purse and heart and hand
Divided we shall surely fall
United we shall stand

And tell your hearts to be as one
And all our veins be free
To fight and rather bleed and die
Than lose our liberty

Come come ye brave Americans
And drink the flowing bowl
May the dear sound of liberty
Sink deep in every soul

Here's a health to North America
And all her noble boys
Her liberty and property
And all that she enjoys

Ann 1776

A SONG ON THE NANTUCKET LADIES

Young damsels all where ever you be
I pray attention give to me
Some braken hints I will lay down
About the girls in Sherbourn town

When eve comes on they dress up neat
And go a-cruising through the streets
To see if they some beaus can find
To suit their fancy and their minds

Skein (?) laces long and frills so neat
And bunnets worked so complete
With their painted cheeks and curled hair
They think to make the young men stare

Their long silk gowns and sleeves so big
You'd think that they had run the rig
With their white kid shoes and silken hose
They look like the devil in their clothes

SONGS THE WHALEMEN SANG

They get the beaus all for to make
A corset board to make them straight
They'll bind it to their waist so tight
And through the streets about from morn till night

Then a few false teeth they're sure to wear
And foretop curls and false hair
And a false heart that'll ne'er prove true
We find it's so it's nothing new

They go to the factory every day
And work twelve months without their pay
And then all for to crown the joke
Why Daniel Dusten is broke

Then Henry Gardner and Peleg West
Then they will do their best
They say that they do what is right
And pay the girls every Saturday night

The girls being few with such ideas
Thinking their master for to please
For six pense a day to work they go
And then they cut a dreadful show

When round Cape Horn their sweethearts go
Then they must have another beau
To wait two years they say they can't
To wait two years they say they shan't

And when their beaus they do come back
Such lamentations they will make
Saying no one has courted me but you my dear
So come along and never fear

Oh says the beaus that never do
I'll never be taken in by you

OF YANKEE MANUFACTURE

If you keep on you'll make me laugh
You can not catch old birds with chaff

But always give the devil his due
There's some will wait 'tis very true
The reason it doth plainly show
They cannot catch another beau

This song was made around Cape Horn
Where most of the young men are gone
Haul down your flag cut down your staff
It is all true you need not laugh

Now to conclude and end my song
There's women tells me I am rong
But if by chance they find it's right
They may sing it from morn till night

Diana 1819

I think that Charles Murphy of Nantucket wrote this. If he did
he would have been nineteen years old at the time.

Sherbourn was the old name for Nantucket village. But who
Daniel Dusten, Henry Gardner and Peleg West were I wouldn't know.
Nor would I know why the girls would work in the factory for a year
without pay.

SPRINGFIELD MOUNTAIN

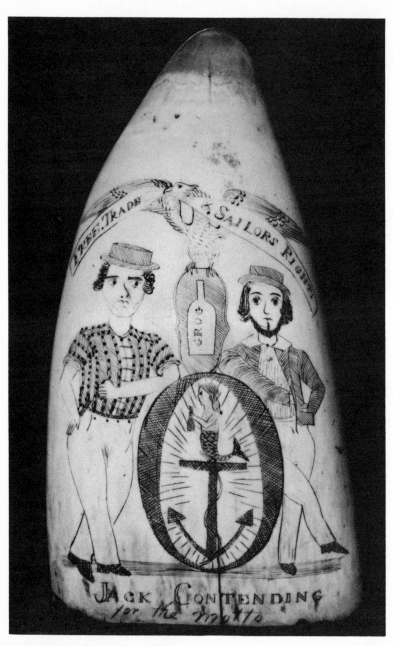

Scrimshaw on a sperm whale's tooth
We'll drink and sing
While foaming billows roll

OF YANKEE MANUFACTURE

Near Springfield Mountain there did dwell
A lovely damsel known full well
Lieutenant Carter's only gal
Her father's joy and named Sal

One day this damsel tript it quick
Down to the stream to berries pick
She hadn't picked but two or three
When her foot slipped and in went she

She uttered an awful yell
When into that swift stream she fell
And then sunk down beneath the wave
Because no hand was near to save

Her lover saw the awful sight
And ran to her with all his might
But when out of the stream he took her
All signs of life had quite forsook her

And from the stream he took her out
And quickly rolled her all about
But when he found her soul had fled
He wrung his hands and cried

And then her lifeless form he bore
Unto her anxious mother's door
Saying Mrs. Carter here you see
All that's left of your Sallie

The awful news spread through her brain
And down she fell nor spoke again
The lover he some poison took
And throwed himself into the brook

And told his ghost to follow arter
His own dear Sal and Mrs. Carter

Braganza 1845

SONGS THE WHALEMEN SANG

The original "Springfield Mountain" is very different from this. There the cause of death was snake bite. Many strange things have happened to "Springfield Mountain" over the years, but surely this parody is one of the strangest.

The true "Springfield Mountain" will be found in many collections, for it is supposed to be one of the earliest, if not indeed, the earliest, of all American folk songs. There are two versions of it in Sharp (1), vol. 2, pp. 166-167. But probably the best of all discussions dealing with the songs is in the various issues of the *Bulletin of the Folksong Society of the Northeast*. This work was reprinted in one volume in 1960 by the American Folklore Society.

If any there be who wish to sing this song — and they need not use my rendering of "Old Hundred" to which the song was probably originally sung — they can make up their own two lines for the ones lacking in the final stanza. Or they can use mine, which go:

> Because you see he sought to be
> With them through all eternity

BLESSED LAND OF LOVE AND LIBERTY

> I've wandered o'er the desert wide
> I've plowed the mighty main
> At every noon and eventide
> I've wandered o'er the plain
> Away from thee away from thee
> I never more will roam
> Blest land of love and liberty
> My own my native home
>
> I've gazed on beauty's bright array
> Upon a far-off shore
> But thought upon the young and gay
> I knew in days of yore
> Away from thee away from thee
> I never more will roam

170

OF YANKEE MANUFACTURE

Blest land of love and liberty
My own my native home

Chili 1839

SONS OF WORTH

Come hail the day ye sons of worth
Which gave your native country birth
All hail the important hour
Let admiration mark the day
When fathers to their sons did say
Be free till time's no more

Columbia's sons have reared a tree
The root and branch are liberty
Expanding far and wide
Refulgent years have rolled away
Since freedom blessed America
Like these two thousand (glide)

When time ruled as George Washington
Not from (affection) he had won
But from the height of care
He left the reins of government
To his successor's management
Quite tired with noise and war

Guide ye gods this reverend sage
Until he's down the steep of age
Then soothe his cares to rest
Yet may his virtues live again
To vindicate the rights of man
Of which we are prospered

John Adams with a finished mind
Columbia's chosen son inclined
To take the important chair

171

The hero takes the reins and guides
America against wind and tides
To shun degrading war

But if to war's terrific sound
He must devote fair freedom's ground
To stain with blood its soil
Then rouse Americans and show
That you can life or wealth bestow
Ere freedom meets a sail

Should Europe's factions once more attempt
To annihilate our government
Or tread upon our shore
Burgoined Cornwallised they would be
Or Arnold like this country flee
Or fall to rise no more

Paulina 1808

This is entitled only "Song" in the journal. Perhaps the proper title should be "Be Free Till Time's No More." I have not yet found a trace of it in print.

JOHN BULL'S EPISTLE

OF YANKEE MANUFACTURE

My Coly Strong my jo Cole what have you been about
You've clearly tried to have me I've fairly found you out
You've proved yourself a traitor you've lied to me you know
You are an arrant coward too oh Coly Strong my jo

Great Coly Strong my jo Cole before the war began
A thousand times you promised that you would be my man
That you'd devide the union you'd split it at a blow
The northern half you'd give to me prince Coly Strong my jo

Great Coly Strong my jo Cole to you our Henry sent
To aid you and your colleagues some project to invent
To put your suitors down Cole your Congress overthrow
You planned but feared to execute oh Coly Strong my jo

Great Coly Strong my jo Cole you were striving to prevent
The inlisting of your subjects to aid your government
Your feeding of my armies with rotten beef and so
And so you have received my thanks for good good Coly Strong my jo

Great Coly Strong my jo Cole when your Convention met
In Hartford famous city to overthrow the States
Great things I did expect Cole would from such wisdom flow
My faith it did in smoke expire poor Coly Strong my jo

Poor Coly Strong my jo Cole when soldiers I did land
Snug on the shores of Castine over which you had command
You dared not long repel me nor fight your Velers no
For which we both despise you poor Coly Strong my jo

Great Coly Strong my jo Cole had all you told proved true
Some royal wished-for title I would have stuck on you

A ribbon or a garter what else I hardly know
My Lord of Essex was the style for Coly Strong my jo

Oh Coly Strong my jo Cole I am constrained to own
That Uncle Sam has flogged me completely got me down
I never would have fought him and that you clearly know
But for your swaggering lang gang oh Coly Strong my jo

Friend Coly Strong my jo Cole advice like this I send
When seated man to pudding be sure your own defend
Lest meddling in your neighborhood you burn your fingers so
That you never will get them well again poor Coly Strong my jo

Poor Coly Strong my jo Cole one thing to me is true
Old Benedict the traitor had more honor far than you
He acted more consistent more independent too
A double traitor he was not oh Coly Strong like you

Herald 1817

The full title of this song in the *Herald* journal is "John Bull's Epistle to My Lord of Essex." But who was "My Lord of Essex?" Could he have been Caleb Strong of Essex County in Massachusetts? If so, Coly Strong was a politician who was very active during the period of the War of 1812, but could he have been as great a traitor as the song indicates?

During the War of 1812 there was a real secessionist movement in New England, as witness the Hartford Convention. The war was very unpopular in the northeast because it was ruining Yankee trade and shipping. And some were willing to risk impressment and even confiscation of vessel and cargo for the chance at the tremendous profits that could result from just one successful voyage.

THE CALIFORNIA SONG

We've formed our band and we are all well manned
To journey afar to the promised land
Where the golden ore is thick in store
On the banks of the Sacramento shore

Chorus

Then ho boys ho to California go
For the mountains bold are covered with gold
On the banks of the Sacramento
Heigh ho away we go
Digging up gold in Frisco

Oh the gold is thar most anywhar
The dig it out with an iron bar
And where it's thick with a spade and pick
They've taken out lumps as big as a brick

Oh don't you cry or heave a sigh
For we'll come back again bye and bye

Don't have a fear or shed a tear
But patiently wait for about two years

We expect our share of the coarsest fare
And sometimes sleep in the open air
Upon the cold ground we will sleep sound
Except when the wolves come howling round

Then we'll roam o'er the dark sea foam
But we'll never forget kind friends at home
For memory kind will bring to mind
The thoughts of those we've left behind

In the days of old the prophet told
Of the city to come all paved with gold
Peradventure they foresaw the day
Now dawning upon California

La Grange 1849

This can only be considered a song the whalemen sang if we assume that there must have been at least one whaleman either in the crew or among the passengers of the *La Grange* when she sailed from Salem for California in 1849. The *La Grange* had been chartered by the California and Salem Mining and Trading Co. to carry its members to California. The song must have reached California with them, and then spread as folk songs do, for it seems to have been collected pretty widely.

William T. Morgan who kept the *La Grange* journal says that the song was composed by Jesse Huchingson (sic) and sung by him and the Barker family at the departure of the expedition. The melody I have used here is from *The Journal of American Folk-Lore*, vol. 35, pp. 361-362 as sung by Mrs. M. M. Moores of Perrysville, Ohio, and I wonder how close it is to the melody that "Jesse Huchingson" composed in 1849. The last four measures of the music have had to be supplied to take care of the extra line of the original chorus.

THE CAPTAIN

Solemn he paced the schooner's deck
I have been where the wild will of Mississippi's tide

176

OF YANKEE MANUFACTURE

Has dashed me on a sawyer
And I have sailed in the thick night
Along the wave-washed edge of ice
By the pitiless coast of Labrador
And I have scraped my keel
O'er coral rocks in   Madagascar seas
And often in my cold and lonely watch
I've heard the warning wash of the lee shore
Speaking in breaking

Aye and I have seen
The whale and swordfish fight beneath my bows
And when they made the deep boil like a pot
Have sung into its vortex and I know
To con my vessel with a sailor's skill
And brave such dangers with a sailor's heart
But never yet upon the stormy wave
Or where the river mixes with the main
Or in the chaffing anchorage of the bay
In all my rough experience of harm
Met I a Methodist meeting house

Cathead or beam or davit has it none
Starboard or larboard gunwale stem or stern
It comes in such a questionable shape
I can not even speak it (lip gib gosy)
And make for Bridgeport there where Stratford Point
Long Beach Fair Weather Island and the bogue
Are safe from such encounters we protest
And Yankee legends long shall tell the tale
That once a Charlestown schooner was beset
Riding at anchor by a meeting house

Lotos 1833

The author of the *Lotos* journal calls this a fragment. But even
as much of it as he could remember seems garbled in places. The Methodist meeting house was probably on a scow which ran afoul of the
Charlestown schooner while she lay at anchor.

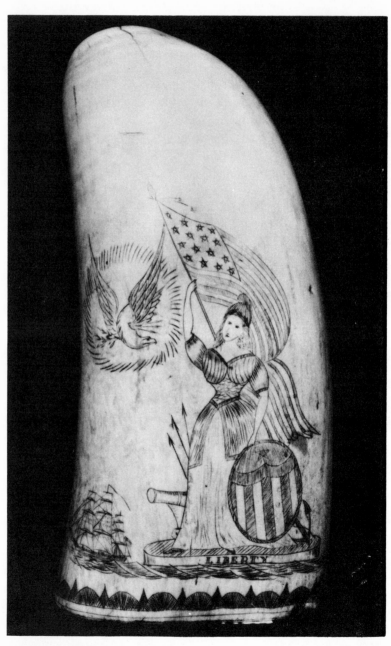

Scrimshaw on a sperm whale's tooth
Come all you jovial sailors
That love your native home

OF YANKEE MANUFACTURE

THE CONFESSION

There's somewhat in my breast father
There's somewhat on my breast
The livelong day I sigh father
At night I can not rest

I can not take my rest father
Though I would fain do so
A weary weight oppresseth me
This weary weight of woe

'Tis not the lack of gold father
No lack of worldly gear
My lands are fair and broad to see
My friends are kind and dear

My kin are kind and true father
They mourn to see my grief
But oh tis not a kinsman's hand
Can give my heart relief

'Tis not that Mary is false father
'Tis not that she's unkind
Though many flatterers swarm around
I know her constant mind

'Tis not her cruel apathy
That chills her laboring breast
'Tis that confounded cucumber
I've et and can't digest

Dartmouth 1836

This little bit of original verse is surely not traditional and probably had no currency whatsoever. Still it is too good to leave out. And there is a lot more good original verse in the *Dartmouth* journal.

VIRTUOUS AMERICA

What is rite or what is rong
In the merry dancing throng
On a summer eve so fair
Dancing in the open air

'Twas on the sixteenth of May
That from my home I strayed away
I saw a man and a lovely maid
Who appeared to be much afraid

I listened to their musick sweet
Until I thought her flesh he'd eat
For he did kiss and hold her tight
At last I screamed with all my might

He left his meat upon the ground
And over the hill he quick did bound
This little maid proved to be
Our honest virtuous America

Euphrasia 1849

THE INDIAN HUNTER

OF YANKEE MANUFACTURE

Let me go to my home in the far distant west
To the scenes of my youth that I like the best
Where the tall cedars are and the bright waters flow
There my parents will greet me white man let me go

Let me go to the spot where the cataract plays
Where oft I have sported in my boyish days
There is my poor mother whose heart will overflow
At the sight of her child oh there let me go

Let me go to the hills and the valleys so fair
Where oft I have breathed my own mountain air
And there through the forest with quiver and bow
I've chased the wild deer oh there let me go

Let me go to my father by whose valiant side
I have sported so oft in the height of my pride
And exulted to conquer the insolent foe
To my father that chieftain oh there let me go

Oh there let me go to my dusky-eyed maid
Who taught me to love beneath the willow's shade
Whose heart like the fawn and pure as the snow
And she loves her dear Indian to her let me go

Oh let me go to my fair forest home
And never again will I wish to roam
And let my body in ashes lie low
To that scene in the forest white man let me go

Marcus 1844

In the *Cortes* journal (1847) there is a version of "The Indian Hunter," almost identical with this one. In spite of that, I had about decided not to include the song here, for it most surely is not a typical folk song, nor does it seem to be a parlor song either. I didn't know what it was, and I still don't, but because I found it in *The Journal of American Folk-Lore,* vol. 35, pp. 375-376, here it is.

SONGS THE WHALEMEN SANG

The melody is from mid-nineteenth century sheet music. 1844 seems early for the feeling of guilt at our treatment of the American Indian that this song so clearly shows.

WILLIE GRAY

My schoolmates now I leave you
I bid you a fond farewell
And the old schoolhouse too I bid adieu
And the scenes I love so well

Chorus

Merrily merrily oh away
O'er the dark blue sea
With a merry heart I leave thee
For a sailor boy to be

I leave thee now for a distant shore
Kind friends when I'm away
When you hear the distant ocean roar
Won't you think of Willie Gray

OF YANKEE MANUFACTURE

Dear father a parting blessing now
O mother a prayer from thee
Kind sister then must I say goodbye
Oh do not weep for me

Brother thy hand before we part
I must I can not stay
The sails are hoisted the ship is ready
They are calling for Willie Gray

Coral 1846

"Willie Gray" was widely sung as a parlor song during the first half of the nineteenth century, and will be found in song books of the period. Many and many a New England boy did leave his schoolmates and the old schoolhouse, too, to become a sailor at twelve or thirteen. One wonders if there is any connection between this song and the better known "Peter Gray."

All From The British Isles

ALL THESE songs are traditional, or were on their way to becoming traditional, and some are very old.

Really it is not strange that so many of the songs that the whalemen sang hail from the British Isles, for the ties between New England and the old country were strong until well into the nineteenth century. Those ties were both cultural and commercial, and for a long time they were stronger than the ties that bound New England to other sections of the United States.

It is true that many of these songs have been collected from the southern mountains, but there, very frequently they are fragmentary and garbled. The reason for that is that in the main, communication between the people of the southern mountains and the mother country had ended by the middle of the eighteenth century, while in New England it had not.

Then, too, there were almost always British seamen on American ships and American seamen on British ships. Indeed, Bill Tilton, my wife's great uncle, who was a New Englander if there ever was one, and a number of whose songs are included in this book, was a chanteyman on British ships for perhaps most of his seafaring years.

THE SHEPHERD'S DAUGHTER

There was a shepherd's daughter
Kept her apron on high
There was a knight swore he'd have his will
As he came riding by

Refrain

Fal de rido etc.

He took her round the middle so small
He laid her on the groun
And when he had his will of her
He helped her up again

Now since you've put my body to shame
Come tell to me your name
In my father's hall they calls me John
In the king's fair court Sweet William

Then he mounted on his milk white steed
And away he did ride
She took her apron in her hand
And she ran by his side

185

When they came to the broken bridge
She held her breast and swam
When they came to the meadow green
She shook her rump and ran

When she came to the king's fair court
She nocked at the ring
There is none so ready as the king himself
To rise and let her in

Good morning good morning my fair maid
Good morning sir said she
There is a man in your fair court
This day has robbed me

Has he robbed you of gold
Or robbed you of (fee)
Oh sir she said he has robbed me
Of the flower of my body
If he is a married man
Hanged he shall be
And if he is a single man
His body I'll give to thee

Joseph Francis 1795

The ink in the *Joseph Francis* journal is badly faded, and to make matters worse the paper has bled in places so I have been forced to guess at a few words. Also I have transposed two lines that were evidently wrongly placed.

This is one of the Child ballads and it has been collected all over the English-speaking world. Versions of it are to be found in many collections. Here are a few: Greenleaf, pp. 35-37; Kidson (1), pp. 19-21; Sharp (2) pp. 6-7; Chappell, pp. 143-144; Williams, pp. 102-103; and JFSS, vol. 3, pp. 222-223, and 280-281; vol. 5, pp. 86-90.

A NEW SONG

Sweet Phillis well met the sun has just set
To yonder cool shades let us repair
All nature is at rest there is none to molest
I have something to say to you there

No no gentle swain entreaties are vain
To persuade me to go you ne'er shall
Night draws on apace I must quit this place
For the dew it's beginning to fall

Believe me coy maid by honor I'm swayed
I mean you no mischief I vow
The oak and the pine their leaves kindly entwine
To shelter love's votrys from harm

All arts I despise my virtue I prize
Though poor I am richer than those
Who lost to all shame will barter their fame
On the purchase of gold or fine clothes

You do me much wrong such thoughts ne'er belong
To the noble and generous breast

I meant but to know if Phillis would go
And let thy love make me blest

If what you now say your heart won't betray
It gives me much pleasure to find him
Unmarried (?) still a stranger to ill
And to wedlock's sweet bondage inclined

Ann 1776

I have played safe with this one by giving it just the title that it has in the *Ann* journal. Its proper title may be "Colin and Phoebe," or "Corydon and Phoebe" or "Sweet Phyllis." Also one of Cecil Sharp's songs called "The Crystal Spring" is undoubtedly related to it.

See Frank Kidson's *Traditional Tunes*, pp. 73-75 for a very interesting discussion of this dialogue type of song. Also Sharp's *English Folk Songs*, vol. 2, pp. 42-43; and Chappell, p. 77.

THE CROPPY BOY

It was early early in the spring
The male birds did whistle and sweetly sing
They so sweetly sing from tree to tree
And the song they sang it was for liberty

It was early on last Thursday night
The Roman cavalry gave me a fright
The Roman cavalry was my downfall
And I was taken by Lord Cornwall

188

It was in his guard house I was laid
It was in his parlor where I was tried
My sentence passed and my spirits low
And to Dungannon I was forced to go

As I was marching through the streets
The drums and fifes did so loudly beat
The drums and fifes did so sweetly play
And to Dungannon we were marched away

As I was marching by my father's door
My brother William stood all on the floor
My poor aging father full grieving and sore
My poor aging mother her hair she tore

When my sister Mary heard of the express
Down stairs she run in her (choring) dress
Five hundred guineas she would a-layed down
To see them marching through Exford town

As I was marching over Exford hill
Who'd blame me if I tarried my fill
I looked behind and I looked before
My poor tender brother I saw no more.

As I was mounting the gallows high
My poor tender brother I saw standing by
My poor tender aged father he did me deny
And the name he gave me was the Croppy boy

I chose the dark I chose the blue
I chose the pink and the orange too
I forsook them all and that I'll ne'er deny
I chose the green and for that I will die

It was in old Ireland this young man died
It was in old Ireland this young man lied

And all you people that are passing by
Say Lord have mercy on the Croppy boy

Galaxy 1827

There is a good version of this song in Patrick Galvin's *Irish Songs of Resistance*, pp. 23-24. There the "Roman Cavalry" is the "Yeoman Cavalry" and "Exford Hill" is "Wexford Hill." Galvin gives a long and detailed explanation of the origin of the term "croppy." The croppies were those who fought against English tyranny.

The melody that I have used here is the first strain of a fiddle tune called "Farewell to Limerick, or the Croppy's Retreat." I found it in a manuscript book of fiddle tunes collected by William Litten on a voyage to China in a vessel of the British India Fleet in the years 1800-1802.

QUEEN OF THE MAY

Now the winter is gone and the summer is come
And the meadows and plantains so gay
I heard a fair maid so sweetly she sung
And her cheeks like the blossoms in May

Young Johnny the plowboy his cheeks like a rose
So cheery he sings to the plow

190

FROM THE BRITISH ISLES

And the blackbird and thrush on every green bush
And the pretty girl a-milking her cow

As I walked through the fields to take the fresh air
And the meadows and plantains so gay
I heard a young damsel so sweetly she sung
And her cheeks like the blossoms in May

I says pretty fair maid oh how come you here
In this meadow this morning so gay
This maid she replied sir to gather me some may
For the trees they are all now in bloom

I says my pretty fair maid shall I tarry with you
In this meadow this morning so gay
This maid she replied are you so innocent
For fear you might lead me astray

I took this fair maid by the lily white hand
And on the green mossy bank set her down
And I planted a kiss on her red ruby lips
And the small birds a-singing all around

And when we arose from the green mossy bank
Through the meadow we wandered away
I had plowed by true love on the green mossy bank
And I plucked her a handful of may

And when we arose she gave me a smile
And thanked me for what I had done
For I planted a kiss on her red ruby lips
For believe me those ne'er would I shun

'Twas early next morning I made her my bride
That the world would have nothing to say
And the bells they shall ring

And the bridesmaids shall sing
And I'll crown her the queen of the may

Bengal 1832

Frank Kidson has a lovely version of this song, more polite than this one, in his *Garland of English Folk Songs.* But one suspects that there may have been versions even much less polite than Ira Poland's in his *Bengal* journal. For the "may" was a pagan survival when maids would wander into the fields for the express purpose of ceasing to be maids.

For other versions of the song see Sharp (2), pp. 120-123; Williams, p. 300. The song has many names among them, "Now the Winter Is Past," and "The Plowboy's Courtship."

THE SHEFFIELD 'PRENTICE BOY

I was brought up in Sheffield not of a high degree
My parents doted on me having no one but me
I rambled about for pleasure just where my pleasure laid
Till I was bound apprentice boy when all my joys were fled

I did not like my master he did not use me well
I formed a strong resolutioin not long with him to stay (dwell?)
Unknown to my parents from him I ran away
I steered my course for London oh cursed be that day

FROM THE BRITISH ISLES

A rich and handsome lady from Holland was there
She offered me great riches to serve her for one year
With long and great persuasions with her I did agree
For to go and live in Holland which proved my destiny

I had not been in Holland scarce a month or two or three
Before that my young mistress grew very fond of me
She says her gold and silver her houses and her land
If I would consent to marry her would be at my command

Oh no dear honored lady I can not wed you both
For I have lately promised and taken a solemn oath
To marry none but Polly your handsome chambermaid
Excuse me my dear mistress she has my heart betrayed

Then in an angry humor away from me she ran
She swore she would be revenged on me before the year was gone
She being so perplexed that she would not be my wife
She swore she would seek some project for to take away my life

So I was a-walking in the merry month of May
The flowers they were springing so delightful and so gay
A gold ring from her finger as she was passing by
She slipped into my pocket now for it I must die

My mistress swore I had robbed her and presently I was brought
Before a grave old judge to answer for my fault
Long time I pleaded innocent but it was of no avail
She swore so hard against me that I was put in jail

It's now my last assizes are drawing to a close
And presently the old judge will on my sentence pass
From a place of close confinement to be taken to a tree
Oh cursed be my mistress for she has ruined me

Come all you young people my wretched fate to see
Don't glory in my downfall ah pray do pity me

Believe me I am quite innocent I'll bid you all adieu
So farewell dearest Polly I will die for the sake of you

Catalpa 1856

This seems to have been a very popular song indeed, and there are many versions of it, but usually the story is pretty much the same and even the melodies to which it was sung all show more or less of a family resemblance. The name was almost always "The Sheffield 'Prentice Boy" but in Kincaid (3), p. 18 there is an interesting version called "Farewell Lovely Polly," and in which the boy was from "Cornish" and not from Sheffield.

See also Creighton (2), pp. 203-206; Sharp (1), vol. 2, pp. 66-69; and JFSS, vol. 1, pp. 200-201, and vol. 2 pp. 169-170. Also JAF, vol. 28, p. 164, and vol. 45, pp 51-53. In both of these latter the boy is from London.

SONG ON COURTSHIP

Yonder stands a handsome lady
Who she is I do not know
Shall I yon and court her for her beauty
What says you madam yes or no

Madam I have gold and silver
Madam I have house and land
Madam I have a world of treasures
And all shall be at your command

What care I for your gold and silver
What care I for your house and land

194

What care I for your world of treasures
All I want is a handsome man

Madam do not count on beauty
Beauty is a flower that will soon decay
The brightest flower in the midst of summer
In the fall it will fade away

The sweetest apple soon is rotten
The hottest love now soon is cold
A young man's word is soon forgotten
The coffin is the end of young and old

A man may drink and not be drunken
A man may fight and not be slaid
A man may court a handsome lady
And be welcome there again

Diana 1819

This old song is known in many versions and there seem to be almost as many titles for it as there are versions. Here are a few: "Twenty Eighteen," "No Sir, No," "Yonder Stands a Handsome Lady," and "Oh No John." Also there are many different melodies for it, and in various versions it is a children's song, a drinking song, and a slightly bawdy song.

See Sharp (1), vol. 2, pp. 279-280 where it is called "Come My Little Roving Sailor." See also Kincaid (1), p. 44; Sharp (2), pp. 154-155; Broadwood, *English County Songs,* pp. 90-91; and JAF, vol. 46, pp. 36-37 where it is called "The Yankee Boys."

WHEN FIRST INTO THIS COUNTRY

When first into this country a stranger I came
There was no one that knew me or asked me my name
I came into this country to tarry for a while
And to leave my dear jewels alone on that isle

A story has been reported as you may have been told
Of three pretty maids from my country I stole
But thievery and robbery oh that I do defy
But these three pretty fair maids I never will deny

The first was pretty Betsy although she was poor
The next was lovely Nancy the maid I adore
The other was black-eyed Susan the girl of my delight
And I'll roll her in my arms on a cold winter night

Her eyes were like two diamonds so bright and so clear
Her cheeks were like the roses the first of the year
Though her cheeks they have been faded this past summer gone
And I fear she's in love with some other young man

I never never knew what it was to be a slave
Till both my parents were dead and laid in their graves
When I was bound apprentice for more than seven long years
My master and my mistress were cruel and severe

My mistress scarce ever spoke a kind word to me
Which caused my mind to wander like waves on the sea
But now I've regained my liberty I'll return to the girl I like
And I'll roll her in my arms on the cold winter nights

Now I am married I live at my ease
I go when I like and I come when I please
I married pretty Susan for she is the girl that I like
And I fold her in my arms on a cold winter night

Cortes 1847

There seem to be two quite different songs called "When First Into This Country." Also there are different titles for the song, as one would expect. In JFSS, vol. 3, p. 309 it is called "The American Stranger." In vol. 5, pp. 50-51 of the same journal it is called "The Irish Stranger." But whatever it is called it is a song of very real charm.

O LOGIE O BUCHAN

O Logie O Buchan O Logie the laird
They have taken away Jemy the delight of the yard
He played on his pipe and his vilot so small
They have taken away Jemy the flower of them all

Chorus
She's a-thinking long lassie though I gang awa'
She's a-thinking long lassie though I gang awa'
The summer is coming cold winter is awa'
I'll come back and see you in spite of them all

Oh Sandy has houses and gear and has kye
A house and lands and silver for by
I would sooner take my ain lad with a staff in his hand
Than him with all his houses and land

My daddy looks sulky my mamy looks sour
They frown upon Jemy because he is poor
I loved them as well as a daughter should do
But not half so well my Jemy as you

I'll sit on my crippy and spin at my wheel
And think on the laddie that I love so weel
We had but one sixpence he broke it in twa
And he gave me one half before he went awa'

Galaxy 1827

In the *Galaxy* journal this song is called only "Scotch Ballad." But "O Logie O Buchan" seems to be its proper title. In *Gems of Scottish Song* it is ascribed to George Halket who died in 1756. There seems no doubt at all but that it had become traditional.

ARAN'S LONELY HOME

If you that has your liberty
I pray you will draw near
A sad and dismal story
I mean to let you hear
While in a distant country
I languish sigh and moan
While I think of the days I spent
In Aran's lonely home

When I was young and in my prime
My age being twenty-one
I had become a servant
Unto a gentleman
I served him true an honest life
And very well 'tis known
But with cruelty he banished me
From Aran's lonely home

The reason why he banished me
I mean to let you know
'Tis true I loved his daughter
She loved me dear also
And she had got a fortune
Her riches I had known

And that is why he banished me
From Aran's lonely home

It was in her father's garden
All in the month of June
A-growing were those flowers
All in their youthful bloom
She said my dearest William
Along with me you may roam
And we'll bid adieu to all our friends
In Aran's lonely home

Unto my sad misfortune
Which proved my overthrow
That very night I gave consent
Along with her to go
The night being bright in moonlight
As we set out alone
A-thinking we might get away
From Aran's lonely home

But when we arrived at Belfast
Just at the break of day
My true love says she'll ready get
Our passage for to pay
Five thousand pounds she counted out
Saying this shall be your own
You will never fret for those you left
In Aran's lonely home

Unto my sad misfortune
Which you shall quickly hear
It was a few hours after
Her father did appear
He marched me away to Omas
In the County of Tyrone
It was there I got transported
From Aran's lonely home

When I received my sentence
It grieved my heart full sore
But the parting with my true love
It grived me ten times more
And when I think upon my chain
And every link a year
Before I can return again
To the arms of my dear

Catalpa 1856

The proper title for this song is "Erin's Lovely Home" and how
it got changed in the *Catalpa* journal is anybody's guess. It seems to have
been a very well-known song. There are versions of it in JFSS, vol. 2,
pp. 167-168, and p. 211; Mackenzie, pp. 117-118; Sharp (2), pp. 124-
125. It is also in Lochlainn, p. 203 where "Omas" is "Omagh Jail," in
the County of Tyrone. And that is probably correct. And for a very
pleasant melody alone see the Hudson manuscript collection of Irish
airs.

FAIR BETSY

Oh Betsy she's a beauty fair
She lately come from Lincolnshire
A hired servant for to be
Which suited Betsy to a high degree

This lady had an only son
And Betsy's beauty his favor won
Oh Betsy's beauty it shone so clear
It drew his heart into a snare

201

SONGS THE WHALEMEN SANG

It was one morning as you shall hear
He says to Betsy you are my dear
I love you dear as I love my life
And I do intend to make you my wife

His mother in a room close by
And hearing what her son did say
Resolved she was in her own mind
To put an end to their design

Early the next morning she arose
And says to Betsy put on your clothes
Into the country you must go
To wait on me a day or two

So Betsy arose put on her clothes
And with her mistress away did go
A ship lay waiting just out of town
And to Virginia fair Betsy's bound

The old woman returning home
Welcome dear mistress said her son
And where is Betsy tell me pray
That she so long behind doth stay

Oh son oh son I plainly see
The love you have for Betsy
But love no more for it is in vain
I have sent her sailing over the main

If Betsy's sailing over the main
May the heavens to my love prove kind
May the heavens to my love prove kind
That she some harbor soon may find

I'd rather see my son lie dead
Than he to Betsy should be wed

Then your desiring you quick shall see
So he pierced his sword through his body

When she saw her son lie dead
With grief and sorrow she hung her head
If my son should be restored again
I'd send for Betsy o'er the main

If Betsy sails upon the main
May the heavens to my love prove kind
She's tender hearted as a dove
She sighs she mourns she dies for love

Cortes 1847

Histed makes the notation for this song in his *Cortes* manuscript that it was sung with great success at Covent Garden. And that suggests that perhaps he copied it directly from a broadside.

Miss Creighton in *Songs and Ballads from Nova Scotia,* pp. 62-63, cites a number of sources where versions of the song may be found. It seems to have been pretty well known. See also Sharp (1), vol. 2, pp. 4-5; JAF, vol. 12, pp. 245-246; vol. 19, pp. 131-132; and vol. 49, pp. 230-231.

I HAD A HANDSOME FORTUNE

I had a handsome fortune
I thought it never'd be sunk
I lost it all by gambling
One night when I got drunk
So early the next morning
My head was wracked with pain
To drive away all sorrow
Then I got drunk again
To drive away all sorrow
Then I got drunk again

And then I went to Germany
All for to get a wife
To leave off all my follies
And lead a happy life
I found a handsome woman
Right happy I was then
But soon it led to strⁱfe sirs
So I got drunk again
But soon it led to strife sirs
So I got drunk again

And then I went to England
All for to learn a trade
I met with two old acquaintances
Two brisk and roving blades
Then we struck up a party
Three jovial fellows then
By drinking free and hearty
So I got drunk again
By drinking brisk and hearty
So I got drunk again

And then I took to gambling
To make up my lost hours
But soon I found by gambling
I lost a great deal more sirs
My friends and my relations
Right happy they were then
But losing all my money
So I got drunk again
But losing all my money
So I got drunk again

It's now I've been in Germany
For more than four long years
And often I've been tipsy

FROM THE BRITISH ISLES

From drinking of strong beer
I've made a resolution
Not to distract my brain
To save my constitution
I'll never get drunk again
To save my constitution
I'll never get drunk again

Cortes 1847

BONAPARTE ON ST. HELENA

Bonaparte is gone
From his wars and his fightings
He has gone to a place
That he never took delight in
He may sit down and tell-o
What great sights he has seen-o
Yet alone he must mourn
On the Isle of St Helena

Where the (Magalene) clouds
Come forth in great splendor
They come forth in crowds
Like the great Alexander
He may sigh to the winds

205

On the great Mount Diana
Yet alone he must mourn
On the Isle of St. Helena

Where the great white-top waves
On the rocks they are dashing
And proud foaming billows
On the shores they are washing
He may sigh to the winds
On the mount of Diana
Yet alone he must mourn
On the Isle of St. Helena

Louisa she mourns
From her husband she is parted
And she dreams when she sleeps
And awakes broken hearted
There is none to console her
Though there's plenty would be with her
Yet alone she mourns
When she thinks on St. Helena

Come all you that have great wealth
Now beware of ambition
Or by some degree or other
You might change your condition
Be steadfast in time-o
What is to come you can not tell-o
Or by chance you might end
On the Isle of St. Helena

George 1829
Galaxy 1827

The version in the *George* journal is slightly more complete than the *Galaxy* version. But both are garbled in places. A. L. Lloyd sent me a very nice version of this song collected in 1904 from one Henry Burstow of Sussex. That version is called "Boney's in St. Helena."

Almost every version of the song that I have found in print has a different title. In Greenleaf, pp. 168-169 it is called "Napoleon the Exile." In JFSS, vol. 2, pp. 88-90 it is called "The Island of St. Helena." And so on. In JAF, vol. 13, p 140 and vol. 35, pp. 358-359 the song is called "Bonaparte on St. Helena," as here.

THE BONNY BUNCH OF ROSES-O

By the borders of the ocean
One morning in the month of June
For to hear those war-like songsters
Their cheerful notes and sweetly tune
I overheard a female talking
Who seemed to be in grief and woe
Conversing with young Bonaparte
Concerning the bonny bunch of roses-o

Then up steps young Napoleon
And takes his mother by the hand
Oh mother dear have patience

SONGS THE WHALEMEN SANG

Until I am able to command
Then I will take an army
Through tremendous dangers I will go
In spite of all the universe
I'll conquer the bonny bunch of roses-o

The first time I saw young Napoleon
Down on his bended knees he fell
He asked the pardon of his father
Which was granted him most mournfully
Saying son I'll take an army
And over the frozen Alps I'll go
And I will conquer Moscow
And return to the bonny bunch of roses-o

So he took five hundred thousand men
And kings likewise to bear his train
He was so well provided for
That he would sweep this world alone
But when he came to Moscow
He was overpowered by the driven snow
When Moscow was a-blazing
So he lost the bonny bunch or roses-o

Oh son don't speak so venturesome
In England there's the hearts of oak
There's England Scotland Ireland
Their vanity was never broke
Oh son think on thy father
On the Isle of St. Helena his body lies low
And you it's soon must follow after
So beware of the bonny bunch of roses-o

Now mother do believe me
Though I lay on my dying bed
If I had lived I would have been clever
But now I droop my fitful head

And while our bones they lie a-mouldering
The weeping willows o'er us grow
The noble deeds of the great Napoleon
Will sting the bonny bunch of roses-o

Cortes 1847

The bonny bunch of roses-o is England. The young Napoleon in the song is the Emperor's son by Marie Louise of Austria about whom not very much is known except that he did not live very long, and most certainly did not follow in his father's footsteps to sting the bonny bunch of roses-o. Like "The Green Linnet" this song is Irish and a lament for a lost cause. There are many versions of it in print, but see particularly Lochlainn, pp. 32-33; Greenleaf, pp. 170-171; Mackenzie, pp. 188-189; and JFSS, vol. 2, p. 276, and vol. 3, pp. 55-57.

BONAPARTE

SONGS THE WHALEMEN SANG

Come all you natives far and near
Come listen to my song and story
Of these few lines you soon shall hear
How soon a man is deprived of glory
Ambition it will have its fling
Fortune backwards it will twiddle
Boni would not be content
Until he was master of the whole world
Fal de ral etc.

He says my wife I will divorce
My dignity is far above her
She gives free scope to all the world
And I will have another lover
Pope and priest I will subdue
They know I am a bold adventurer
And like St. Ruth my name shall rue
Since I became the royal emperor

He says I will rise above the moon
And climb the air through snow and thunder
And rise up like an air balloon
And cause all nations for to wonder
There is no man can turn my head
I can tear down the walls of China
Not dreaming of the countermand
For to embark for St. Helena

Boni was a hero bold
He was the terror of the whole dominion
He would form a plan and raise a scheme
That would bring thousands to their ruin
For peace with Briton he would not make
He says their wooden walls I will shiver
Old England's Isle I'll over and take
And immortalize my name forever

To Waterloo his troops he drew
He says my boys I will never surrender
All nations we will rule and take
Like the glorious Alexander
But Wellington he took the field
The British boys they thought to baffed them
At last poor Boni was forced to yield
And run on board the Baldorphan

Now mark the fate of this great man
He thought all nations to subdue them
He would form a plan and raise a scheme
That would bring thousands to their ruin
It is now my darling wife he cried
Fairer than Eland or bright Dianah
It is for you I lament for life
Within my bounds on St. Helena

L. C. Richmond 1834

In the *Richmond* journal the title is spelled "Bonny Parte." The nearest thing that I have found to this particular "Bonaparte" is one called "Boney's Lament" in Lucy Broadwood's *English Traditional Songs and Carols,* pp. 34-35.

In very many of the ballets in the seamen's journals a chorus or refrain will only be indicated as it is here, I suppose because it was taken for granted that everyone knew the chorus anyway. I particularly like the spelling "Baldorphan" for "Bellerophon," the British war vessel that carried Napoleon to his final imprisonment on St. Helena. The melody I have used is in JFSS, vol. 1, pp. 166-167, and it fits. There the song is called "Boney's Lamentation." The refrain is probably only four measures long.

THE GREEN LINNET

Curiosity bore a young native of Erin
To view the gay banks of the Rhine
There an empress he saw and the robes she was wearing
All over with diamonds did shine
A goddess in splendor was never yet seen
To equal this fair one so mild and serene
In soft murmur she says my linnet so green
Are you gone will I never see you more

The cold lofty Alps you freely went over
Which nature had placed in your way
That Marengo Saloney around you did hover
And Paris rejoiced the next day
It grieves me the hardships you did undergo
The mountains you travelled all covered with snow
The balance of power your courage laid low
You are gone I will never see you more

The crowned heads of Europe when you were in splendor
Fain they would have had you submit
But the goddess of freedom soon bid them surrender
And lower the standard to your wit
Old Frederick's colors to France you did bring
Yet his offspring found shelter under your wing

FROM THE BRITISH ISLES

That year in Virginia you sweetly did sing
Are you gone will I never see you more

The numbers of men that were eager to slay you
Their malice you viewed with a smile
Their gold through all Europe they sowed to betray you
And they joined the Mamelukes on the Nile
Like ravens for blood their vile passions did burn
The orphans they slew and caused widows to mourn
They say my linnet's gone and n'er will return
He is gone will I never see him more

When the trumpet of war the grand blast was sounding
You marched to the north with good will
To relieve the poor slaves in their vile sack clothing
You used your exertion and skill
You spread out the wings of your envied train
When tyrants great Caesar's old nest set in flames
Their own subjects they caused to eat herbs on the plains
Are you gone will I never see you more

In great Waterloo where numbers laid sprawling
In every field high or low
Fame on her trumpets the Frenchmen were calling
Fresh laurels to place on her brow
Usurpers did tremble to hear that loud call
The third babe's fine new buildings did fall
The Spaniards their fleet in the harbor did call
Are you gone will I never see you more

I'll roam the deserts of wild Abyssinia
And yet find no cure for my pain
I'll go and enquire on the Isle of St. Helena
No there we will whisper in vain
Now tell me ye critics now tell me in time
The nation I'll range my sweet linnet to find

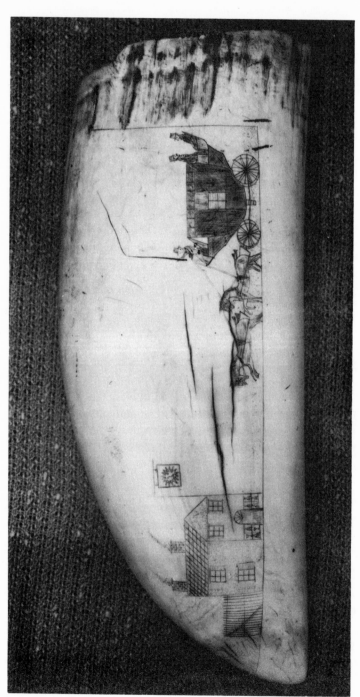

Scrimshaw on a sperm whale's tooth
He has been a gallant voyage
And has lately come on shore

Was he slain at Waterloo or at Elba on the Rhine
If he was I shall never see him more

Cortes 1847

Here is another true lament for the loss of the great hope that the Irish had had in Napoleon. And here again, as in "The Bonny Bunch of Roses-O" we see the use of symbolism, for, the green linnet is Napoleon.

No matter how twisted the history may be in places, the story of Napoleon is here, and when the green linnet went down to final defeat the hopes of many went with him. There is little recognition of the fact in these songs that the Corsican himself might have betrayed the high hope that had been placed in him.

This song does not seem to have had nearly as much currency as "The Bonny Bunch of Roses-O" or "Bonaparte on St. Helena." There is a version of it in JFSS, vol. 7, pp. 151-152; and also see Barry (2), p. 66. And there is a quite different melody for it in Hudson's manuscript collection. See also JAF, vol. 67, pp. 129-130.

ONE NIGHT SAD AND LANGUID

One night sad and languid I went to my bed
And had scarcely reclined on my pillow
When a vision surprising came into my head
And methought I was crossing the billow
I thought as my vessel sped over the deep
I beheld that rude rock that grows craggy and steep
Where the willow (the willow) is now seen to weep
O'er the grave of the once famed Napoleon

Methought as my vessel drew near to the land
I beheld clad in green his bold figure
With the trumpet of fame he had clasped in his hand
On his brow there shone valor and rigor
He says noble stranger you have ventured to me
From that land of your fathers who boast they are free
If so then a tale I will tell unto thee
'Tis concerning that once famed Napoleon

You remember the day so immortal he cried
When we crossed o'er the Alps famed in story
With the legions of France whose sons were my pride
As I marched them to honor and glory
On the fields of Marien lo I tyrany hurled
Where the banners of France were to me first unfurled
As a standard of liberty all over the world
And a signal of fame cried Napoleon

Like a hero I've borne both the heat and the cold
I have marched to the trumpet and cymbal
But by dark deeds of treachery I now have been sold
Though monarchs before me have trembled
Ye princes and rulers whose station ye bemean
Like scorpions ye spit forth venem and spleen
But liberty all over the world shall be seen
As I woke from my dream cried Napoleon

Cortes 1847

This is another of the songs that show so clearly the strength of the Napoleonic myth. The line, "From that land of your fathers who boast they are free," seems to indicate that this particular Napoleon song is American. But most of the song has the feel of an Irish lament.

A FARMER'S BOY

FROM THE BRITISH ISLES

The sun had sunk behind the hills
Across yon dreary moor
When wet and cold a boy there came
Up to a farmer's door
Can you tell me said he if any there be
Who would like to give employ
For to plow and to sow
For to reap and mow
For to be a farmer's boy

My father's dead my mother's left
With four poor children small
And what's the worse for my mother dear
I'm the eldest of them all
But though I am young I'll do all I can
If I can get employ
For to plow for to sow
For to reap for to mow
For to be a farmer's boy

But if no boy by chance you want
One favor I've to ask
That you'll shelter me till the dawn of day
From the cold and wintry blast
And at break of day I will trudge away
Elsewhere to seek employ
For to plow for to sow
For to reap for to mow
For to be a farmer's boy

The farmer's wife cries try the boy
Let him no farther seek
Oh do dear papa the daughter cries
While the tears run down her cheek
For those who will work 'tis hard to want
Or to wander for employ
For to plow for to sow

For to reap for to mow
For to be a farmer's boy

The farmer's boy he grew a man
The good old farmer died
He left the lad all that he had
With his daughter for his bride
The boy that now a farmer is
Oft thinks and smiles with joy
As he thinks of the day
When he passed that way
For to be a farmer's boy

Elizabeth 1845
Cortes 1847

This seems to be one of those songs so very well known that it has become almost completely standardized. Another example of an almost completely standardized folk song is "The Little Brown Jug." But in the latter song the melody also has become standardized, while there are several good tunes for "A Farmer's Boy."

It is a nice little song, and those who would like to compare some of the melodies may see the *Penguin Song Book,* pp. 118-119; Creighton (2), p. 158; Kidson (1), pp. 63-66; Kidson (2), pp. 100-101; and Broadwood (1), pp. 120-121, and also pp. 134-135. Also JAF, vol. 51, pp. 38-39, and vol. 52, pp. 37.

THE COUNTY OF TYRONE

My father oft told me he n'er would control me
He'd make me a draper if I'd stay at home
But I had a notion of a higher promotion
To travel some county as well as Tyrone

It was at a variance I parted from my parents
Little they knew the road I had gone
But I thank my instructor and kind conductor
Who landed me safe from the County Tyrone

I travelled all round till I came to Dover
Where the pretty girls there surrounded me with their charms
And all of them asked me if I was not for Glasbury
And had travelled the country all round from Tyrone

When I came to Dover there I fell a-weaving
I courted a fair maid for a wife for my own
With great apprehension she quickly made mention
Saying what is your character you brought from Tyrone

As for my character you never need mind it
I never was courted or married to one
She swore by her conscience she would run all chances
And travel with me to the County Tyrone

Early the next morning as the sun was rising
We went from Kyde Cock to the three mile stone
The guard he pursued us but never could view us
I wished in my heart I'd my love in Tyrone

As we were a-walking and pleasantly talking
We met an old man who was all alone

He told them he met us and where they could get us
And that we were talking of the County Tyrone

This ended their trouble and their steps they doubled
They swore if they got me they'd break all my bones
They swore if they got me a prisoner they'd make me
Transport me to Antigua or hang me in Tyrone

A channel being nigh where vessels did lie
To them my whole story I quickly made known
They threw a plank to us and on board they drew us
They told us the vessel was bound to Tyrone

And when we came to my own native country
To my parents the whole story I quickly made known
Five hundred pounds they gave us and if that would not do us
They crown us with honor in the County Tyrone

These two lived together with joy and great pleasure
And if you would see them you must go to Tyrone
My love's name to finish is sweet Pennie Eunis
And mine is bold Magnie Eunis of the County Tyrone

Cortes 1847

There is a version of this song in the National Library of Ireland. There it is No. 153 of Sam Henry's *Songs of the People*, an unpublished collection of songs, partly in manuscript and partly as newspaper clippings, collected by the late Sam Henry in the North of Ireland for the *Northern Constitution* of Coleraine.

In Sam Henry's version the town is not Dover but Newry. He says that vessels of about seventy tons used to run between Newry and the County Tyrone. The channel "where vessels did Lie," must be the Lough Neagh Canal which was built about 1745.

IN DAYS WHEN WE WENT GIPSYING

In days when we went gipsying
A long time ago

The lads and lasses in their best
Were dressed from top to toe
We danced and sung the jocund song
Upon the village green
And naught but mirth and jolity
Around us could be seen

Chorus

And so we passed a pleasant time
Nor thought of care or woe
In days when we went gipsying
A long time ago

All hearts were light and eyes were bright
And nature's face was gay
The trees their leafy branches spread
And perfume filled the (day)
'Twas there we heard the cuckoo's note
Steal softly through the air
And every spot around us seemed
Most beautiful and fair

We filled a glass to every lass
And friends we loved most dear
And wished them many a happy day
And many a happy year
To friends away we turned our thoughts
With feelings kind and free
But oh we wished them there with us
Beneath the old oak tree

Cortes 1847

A notation following this song in the *Cortes* manuscript reads, "Ship Splendid, July 3rd 1847," from which I take it that Histed got this song during a gam with that vessel. It is a nice song, too. I have found it only in Williams' *Folk-Songs of the Upper Thames*, p. 69.

SONGS THE WHALEMEN SANG

It is sad that Williams never got around to gathering the music for his songs which he says he had intended to do. The last stanza of the Williams version is quite different from Histed's.

RINORDINE

One evening as I rambled two miles below Pomsey
I met a farmer's daughter all on the mountains high
I said my pretty maiden your beauty shines most clear
And upon these lonely mountains I'm glad to meet you here

She said young man be civil my company forsake
For to my great opinion I fear you are a rake
And if my parents should know my life they would destroy
For keeping of your company upon the mountains high

I said my dear I am no rake and brought up in (and) train
And looking out for concealments all in the judge's name
Your beauty has ensnared me I can not pass you by
And with my gun I'll guard you all in the mountains high

This pretty little thing she fell into amaze
With her eyes as bright as nature on me she did gaze
Her cherry cheeks and ruby lips they lost their former dye
And then she fell into my arms all on the mountain high

I had but kissed her once or twice till she came to again
She modestly then asked me pray sir what is your name
If you go to yonder forest my castle you will find
Wrote in ancient history my name is Rinordine

I said my pretty fair maiden don't let your parents know
For if you do they'll prove my ruin and fatal overthrow
But when you come to look for me perhaps you'll not me find
But I'll be in my castle and call for Rinordine.

222

FROM THE BRITISH ISLES

Come all ye pretty fair maidens a warning take by me
And be sure you reject night walking and shun bad company
Foe if you don't you'll surely rue until the day you die
And beware of meeting Rinordine all on the mountains high

Sharon 1845

Miss Creighton in *Maritime Folk Songs,* pp. 112-113, places "Rinordine" among the songs of the supernatural. However, it is hard to find any trace of the supernatural in this particular version. See also Mackenzie, pp. 102-103, and JAF, vol. 18, p. 322.

BEHIND THE GREEN BUSH

On a primrose bank by a murmuring stream
Pastora sat singing and I was her theme
Whilst charmed with her beauty behind a green bush
I listened to her sweet tale with a blush

Of all the young shepherds that pipe and reed
'Tis Damon alone I can fancy indeed
I tell him I value him not of a rush
Yet foreby I love him or why do I blush

When I went to a grove at the top of a hill
It was last May I remember it still
He brought me a nest of young linnets quite flush
And I his kind present received with a blush

Whenever he meets me he'll simper and smile
I seem as if I did not observe him the while
He offers to kiss me and I give him a push
Why can't you be easy I cry with a blush

One Sunday he came to entreat me to walk
'Twas down in a meadow of love was our talk
He called me his dearest pray Damon to hush
There's somebody coming I cry with a blush

My mother she chides me when I mention a swain
Forbids me to go to the meadow again
But brave for his sake I venture a brush
For love him I do I confess with a blush

Thus warbled my fair and my heart leaped for joy
Though little she thought her Damon was nigh
But chancing to spy on me behind a green bush
She ended her song and rose with a blush

Two Brothers 1768

At first I thought that this pretty little song was a version of "The Green Bushes," but it evidently is not. It did have a title in the *Two Brothers* journal, "The Shepherd's Resolution," but that has been crossed out, so until I find its proper title "Behind the Green Bush" will have to do.

REILY'S JAILED

Come rise up William Reily and come along with me
I mean for to go with you and leave this country
I'll forsake my father's dwellings his houses and rich land
And go along with you my love your dear Coleen Bawn

Over lofty hills and mountains along the lonesome dales
Through shady groves and fountains rich meadows and sweet vales

We climbed the rugged woods and sped on the silent lawn
But I was overtaken with my dear Coleen Bawn

They hurried me to prison and my hands and feet they bound
Confined me like a murderer with chains unto the ground
But this hard cruel treatment most cheerfully I'll stand
Ten thousand deaths I'd suffer for my dearest Coleen Bawn

Sharon 1845

"Reily's Jailed" is called usually "Willy Rily" or "The Courtship of Willy Rily." Miss Creighton in *Songs and Ballads from Nova Scotia,* pp. 152-162, gives seventy-eight stanzas of the song and says that there are more. But strangely enough, the little three stanza version in the *Sharon* journal is a complete song, and entire in itself. Seventy-eight stanzas versus three! It is this process of drastic shortening, but not fragmentation, by which ballads sometimes turned into lyrics. See also Greenleaf, pp. 184-186.

The melody I have used here is from *The Journal of American Folk-Lore,* vol. 24, p. 340.

THE FIRST TIME I SAW MY LOVE

The first time I saw my love happy was I
For I knew not what love was nor how to deny
For I made too much freedom of my love's company
My generous lover you are welcome to me

Happy is the maid that n'er loved a man
She is free of all sorrows that we understand
She is free of all sorrows and sad misery
Oh my generous lover you are welcome to me

My friends and relations they angry were all
For to make free with you in younder fine hall
But my friends and relations they angry may be
My generous lover you are welcome to me

Scrimshaw on a sperm whale's tooth
Fanny Blair is a girl of eleven years old

FROM THE BRITISH ISLES

And it's farewell my lassie since I must away
For I in this country no longer can stay
So it's keep your mind easy love keep your mind free
And let no other man be sharers but me

Oh this innocent creature she stood on the ground
With her red rosy cheeks and the tears falling down
Saying Jimmy dear Jimmy you're the first that wooed me
My generous lover you are welcome to me

Catalpa 1856

I have not been able to find this very pretty little song from the *Catalpa* journal. A. L. Lloyd suggested that it might be Irish, but on the other hand, R. J. Hayes of the National Library of Ireland thinks it is probably English. Perhaps "The First Time I Saw My Love," is not the proper title, but that doesn't help much either.

Somehow it is vaguely reminiscent of "O Logie, O Buchan," but as it stands here, it most certainly is not that song.

THE SHEPHERD'S LAMENT

One morning (one morning) one morning in May
The fields was adorning with costly display
I chanced for to hear as I walked by a grove
A shepyard laymanting for the loss of his love

Was there ever a man in so happy a state
As I with my flower my flower so great
I stepped to my fair flower and to her did say
For to make us both happy it wants but one day

Oh no says the fair one that day is not come
Oh no saucy shepherd our years are too young
First I go to sarvis and when I return
Then we will be married and our love carried on

Will you go to sarvis and leave me to die
Oh yes saucy shepherd I'll tell you for why

227

For to marry so early I think it's not fit
For some years will espier both our sumptence and wit

As fortune would have it to sarvis she went
To wait on a lady it was her intent
She chanced for to meet with a lady so gay
Who clothed the fair flower with costly array

In a week or two after a letter he sent
In two or three words for to know her intent
She wrote in a word that she lived such a life
That she never did intend to be a gardiner's wife

Vaughn 1767

This song has no title in the *Vaughn* journal. I have called it "The Shepherd's Lament" as seeming to fit. It seems to be related to several songs, among them the "Green Bushes."

I have corrected most of the spelling but I felt that I had to leave the "shepyard laymanting" for the loss of his love. And in the fourth stanza, "espier both our sumptence and wit" means to advance both our substance and wit. Welcome Tilton, my wife's grandfather used to use both those words with exactly that meaning.

See Henry No. 30; Broadwood (2) pp. 128-129; Creighton (3) p. 82.

WOMEN LOVE KISSING AS WELL AS THE MEN

A slave to a fair from my childhood I've been
Before a soft down appeared on my chin
And 'tis from experience all manners are known
I've found 'em all kind from Clarinda to Joan
Young Cloe was wanton but scruples she had
I wooed her so closely she yielded egad
It's now you'll be constant she whimpered and cried
I knew what I thought so I replied

My dear can you doubt it I kissed her again
For women love kissing as well as the men

Kind Celia devoted kind lectures to me
She wondered what pleasures in kissing could be
I pressed her to try it and then speak her mind
She made a sweet proof and grew instantly kind
Then answered me softly I'll try it again
For women love kissing as well as the men

That women are cruel is all a mistake
For every young female at heart is a rake
This conduct a lover the damsel'll secure
Stick close to her lips she's unalterably yours
And search through the sex I'll lay twenty to ten
All women love kissing as well as the men

Two Brothers 1768

I have no information on this song. It probably should consist of four six line stanzas, although it is hard to tell from its format in the journal. If so, two lines were omitted from the second stanza.

FANNY BLAIR

SONGS THE WHALEMEN SANG

Come all you young men and maidens where ever you may be
Beware of false swearing and sad perjury
For it is by a false woman I am wounded so soon
And you see how I am cut down in the height of my bloom

It was last Monday morning I lay in my bed
A young friend came to me and unto me said
Rise up Dennis Higgins and flee you elsewhere
For they're now down against you for the young Fanny Blair

Fanny Blair is a girl of eleven years old
And if I was a-dying the truth I'd unfold
It's I never had dealings with her in my time
And it's I have to die for another man's crime

On the day of the trial squire Vernon was there
And it's on the green table he handed Fanny Blair
And the oath that she swore I am ashamed to tell
And the judge spoke up quickly you have told it well

Dennis Higgins of Branfield whither art thou flown
That you are a poor prisoner condemned and alone
If John O'Neil of Shane's Castle only was here
In spite of (Dawson) n'er known he'd soon set you clear

On the day that young Higgins was condemned to die
The people rose up with a murmuring cry
Go catch her and crop her she's a perjuring whore
Young Dennis is innocent we are very sure

One thing yet remaining I ask you my friends
To wake me in Branfield amongst my dear friends
Bring my body to lie in Merrylee mold
And I hope that great God will pardon my soul

Java 1839

One thing seems quite sure about this strange song. That is that it tells the story of an actual crime, and quite possibly poor Dennis Higgins was innocent.

In the Centenary Edition of Cecil Sharp's *English Folk Songs* there is a version of the song where the young man's name is Thomas Hogan. Squire Vernon is the same in both versions, though Sharp has no John O'Neil, nor Shane's Castle. But this is the big difference in Sharp's version: his Fanny Blair is not eleven years old but eighteen, thus the whole "Lolita-like" quality of the song is lost. And one wonders if Sharp did not change the girl's age to fit some late Victorian concept of propriety. But the tune Sharp uses is strange and beautiful, A. L. Lloyd calls it among the most remarkable of British airs.

There is also a version of the song in Loraine Wyman's *Twenty Kentucky Mountain Songs*. See also JAF, vol. 30, pp. 343-344. There Fanny Blair is "scarce" eleven years old.

Parlor Songs That Went To Sea

THESE parlor songs — "art songs" is really the proper term — were mostly written and composed to meet the standards of cultivated society of the period. Young ladies and gentlemen of refinement would sing these songs to the accompaniment of piano or parlor organ, who would not dream of sullying their lips with a "common song." And "common song" meant folk song. The whalemen made no such distinction, that is most of them did not. To them a song was a song, and if they liked it they sang it.

So these parlor songs that went to sea can not be called folk songs in the generally accepted meaning of the term. But the definition of folk song is difficult, because it means so many different things to different people. All of these parlor songs seem to have been transmitted by oral tradition, many of them show considerable change, and some of them were surely on the way to becoming "true" folk songs. Perhaps the nearest we can come to it is Wilgus' definition of what the Germans call *volktümliches Lied*, that is, folk-transmitted song, not "true" folk song.

But they must be included here because the whalemen sang them. And some of them are well worthy of being included.

SILVERY MOON

I strayed from my cot at the close of the day
To muse on the beauties of June
By the Jessamin shade I espied a fair maid
And she sadly complained to the moon
Roll on silver moon guide the traveller his way
While the nightingale sings in tune
For never never more with my love will I stray
By the sweet silver light of the moon

It's a hart on the mountain my lover was brave
So handsome so manly to view

So kind and sincere and he loved me so dear
Edwin my love was more true
But now he is dead and the youth once so gay
Is cut down like a rose in full bloom
But he silently sleeps while I'm thus left to weep
By the sweet silver light of the moon

But his grave I'll seek out until morning appears
And weep for my lover so brave
I'll embrace the cold earth and bedew with my tears
The flowers that bloom o'er his grave
Oh never again can my heart throb with joy
My lost one I hope to meet soon
And kind friends will weep o'er the grave where we sleep
By the sweet silver light of the moon

Euphrasia 1849
Cortes 1847

In the *Euphrasia* version of "Silvery Moon" there is this notation,
"Roll on silver moon, etc." which may indicate that the last four lines
of the first stanza were sung as a chorus or refrain. The opening lines
of the *Cortes* version read:

As I went to my cot at the close of the day
'Twas about the beginning of June.

I doubt if many would call this a true folk song, but Williams
includes it in his *Folk Songs of the Upper Thames*, p. 128.

WILLIE'S ON THE DARK BLUE SEA

234

My Willie's on the dark blue sea
He's gone far o'er the main
And many a weary day will pass
E'er he'll come home again

Then blow gentle winds o'er the dark blue sea
Bid the storm king stay his hand
And bring my Willie back to me
To his own dear native land

I love my Willie best of all
He ever was true to me
But lonesome dreary are the hours
Since first he went to sea

Blow gentle winds o'er the dark blue sea
Bid the storm king stay his hand
And bring my Willie back to me
To his own dear native land

There's danger on the waters now
I hear the bland hills cry
And moaning voices seem to speak
From out the cloudy sky

I see the vivid lightning flash
And hark the thunder's roar
Oh father save my Willie from
The storm king's mighty power

And as she spoke the lightning ceased
Hushed was the thunder's roar

And Willie clasped her in his arms
To roam the seas no more

Euphrasia 1849

This is a parlor song, but the type of parlor song that could easily become traditional. It was a Miss Elmira Higgins who recorded this version of it in the *Euphrasia* journal. For a fragment of the song that does truly seem to be traditional, see JAF, vol. 52, p. 27.

THE BANKS OF BANNA

Shepherds have you seen my love
Have you seen my Anna
Pride of every shady grove
Upon the banks of Banna

I for her my home forsook
Near yon mighty mountain
Left my flock my pipe my hook
Greenwood shade and fountain

Never shall I see them more
Until her returning
All the joys of life are o'er
From gladness turned to mourning

Whither is my charmer flown
Shepherds tell who've seen her
Ah woe's me perhaps she's gone
For ever and forever

Joseph Francis 1795

This little song with the terrific range is to be found in James Wilson's *The Musical Cyclopedia,* p. 20. The song is also in Chappell, and he raises the question as to whether or not it is traditional.

THE MAID OF ERIN

My thoughts delight to wander upon a distant shore
Where lovely fair and tender is she whom I adore
May heaven its blessings sparing on her bestow them free
The lovely maid of Erin who sweetly sang to me

Had fortune fixed my station in some propitious hour
The monarch of a nation endowed with wealth and power
That wealth and power sharing my peerless queen would be
The lovely maid of Erin who sweetly sang to me

Although the restless ocean may long between us roar
Yet while my heart has motion she'll lodge within its core
For artless and endearing and mild and young is she
The lovely maid of Erin who sweetly sang to me

When fate gives intimation that my last hour is nigh
With placid resignation I'll lay me down and die
Fond hopes me cheering that in heaven I'll see
The lovely maid of Erin who sweetly sang to me

Cortes 1847

ADIEU MY NATIVE LAND

Adieu my native land adieu
The vessel spreads her swelling sails
Perhaps I never more may view

Your fertile fields and flowery dales
Delusive hope no more can charm
Far from the faithless maid I roam
Unfriended leave my native shore
Unpitied leave my peaceful home

Farewell dear village oh farewell
Soft on the gale thy murmur dies
I hear the solemn evening bell
Thy spire yet glads my aching eyes
Though frequent falls the dazzling tear
I scorn to shrink from fate's decree
Yet think not cruel maid that e'er
I'll breathe another sigh for thee

In vain through shades of frowning night
My eyes thy rocky coast explore
Deep sinks the fiery orb of light
I view thy beacon now no more
Rise billows rise blow hollow winds
Nor night nor storm nor death I fear
Unfriended bear me hence to find
The peace that fate denies me here

Cortes 1847

I found the melody for this song in vol. 3 of the *Franklin Square Song Collections,* and there the words of the song differ from the version in the *Cortes* manuscript. The song seems to have been well on the way to becoming traditional.

ANGELS WHISPER

A baby was sleeping its mother was weeping
For her husband was far o'er the raging sea
And the tempest was swelling round the fisherman's dwelling
And she cried Dermot darling oh come back to me

The beads while she numbered the baby still slumbered
And smiled in her face while she bended her knee
Oh blessed be that warning my child's face adorning
For I know that the angels are whispering to thee

And while they are keeping bright watch o'er thy sleeping
I pray to them softly my baby with me
And say thou would rather they'd watch o'er thy father
For I know that the angels are whispering to thee

The dawn of the morning saw Dermot returning
And the wife wept with joy the babe's father to see
And closely caressing the child with a blessing
Said I knew that the angels were whispering to thee

Cortes 1847

The words of this song were written by Samuel Lover. The melody is an old Irish air, "Mary Do You Fancy Me." But the song also sings very well to the tune of "Molly Malone."

THE BRIDE'S FAREWELL

Farewell mother tears now streaming
Down thy pale and tender cheek
While I've gems and roses beaming
Scarce the hard farewell can speak

Farewell mother now I leave thee
Hope and fear my bosom swells
One to trust who may deceive me
Farewell mother fare thee well

Farewell father thou art smiling
Yet there's creases on thy brow
Winning me from that beguiling
Tenderness to which I go

Farewell father thou dids't bless me
Ere my lips thy name could tell
He may wound who can caress me
Father guardian fare thee well

Carving on whale or walrus ivory
There was something in your glances
Put a summer in our fancies

Farewell sister thou art twining
Round me an affection deep
Wishing joy in near devining
Why a blessed bride should weep

Farewell brave and gentle brother
Thou art more dear than words can tell
Father mother sister brother
All beloved ones fare thee well

Fortune 1840

The only printed version of this song that I have found is in vol. 8 of the *Franklin Square Song Collection*. It has a sprightly, happy little melody which somehow goes very well with the semi-sad words of the song.

THE DYING SOLDIER

He was young for years not twenty
O'er his sunny head had flown
Yet far away from home and loved ones

243

He was dying all alone
Strangers smoothed his dying pillow
Wiped the death damp from his brow
But a gentle mother's memory
Lingered round the soldier now

Restlessly his blue eyes wandered
Over each one standing by
But he sadly murmured mother
Come and see me ere I die
Oh mother come for I am dying
And my heart throbs with pain
Come and bring sweet sister with you
She will make me well again

Then they smoothed the sunny ringlets
Off the brow so pale and fair
Pressed the lids down o'er the blue orbs
And then left him sleeping there
Where the sunny balmy breezes
With the gentle flowerettes toy
When the stars look down at midnight
Angels guard the soldier boy

Lexington 1853

This sad and rather commonplace song of the Mexican War, through
a metamorphosis common enough with folk song, seems to have changed
into "The Dying Ranger" a better song and truly traditional. Here are
two stanzas from that latter song which I learned in Florida many years
ago:

The sun was sinking in the west
And fell with lingering rays
Through the branches of the forest
Where the wounded ranger lay
'Neath the shade of a palmetto
And the sunset's silvery sky

Far away from his home in Texas
We layed him down to die

Our country was invaded
They called for volunteers
She threw her arms around me
Then bursting into tears
She says my darling brother
Drive those traitors from our door
My heart may need your presence
But our country needs you more

For versions of "The Dying Soldier" and "The Dying Ranger" see
JAF, vol. 45, pp. 164-165, and vol. 46, pp. 27-28.

GENETTE AND GENOE

You are going far away from your poor Genette
There is no one left to love me now and you too may forget
Yet my heart it shall be with you wherever you do go
Can you look me in the face my love and say the same Genoe

With your lovely jacket on and your beautiful cockade
You will be forgetting all those promises you made
With your gun upon your shoulder and your bayonet by your side
You'll be courting some fair lady and making her your bride

When glory leads the way and you're rushing madly on
Not thinking if they kill you my happiness is done
And should victory crown the day and some general you'll be
Though I'd be proud to hear of this what would become of me

If I was the king of France or what's more the Pope of Rome
I'd have no fighting man abroad or weeping maid at home
But all the world would be at peace and man maintain his right
And them that made the guards 'd be the only ones left out

Euphrasia 1849
Minerva Smythe 1852

The proper title for this song is "Jeanette and Jeannot." It was written by Charles Glover and Charles Jeffreys in the latter part of the eighteenth century and for many years was a remarkably popular parlor song. However, it does seem to have become truly traditional with seamen. Both the *Euphrasia* and *Minerva Smythe* versions show changes from the original other than mere spelling. In the *Euphrasia* journal it is entitled, "The Conscript's Departure."

MARY'S DREAM

The moon had climbed the highest hill
That rises o'er the source of Dee
And o'er the eastern summit shed
Its silver light on tower and tree
When Mary laid her down to sleep
Her thoughts on Sandy far at sea
When soft and low a voice she heard
Saying Mary Mary do not weep
Oh do not weep for me

She from her pillow gently raised
To hearken to ask who there might be
She saw young Sandy shivering stand
With pallid cheek and hollow eye
Oh Mary dear cold is my clay
It lies beneath the stormy sea

Far far from here I sleep in death
Oh Mary Mary do not weep
Oh do not weep for me

Three stormy nights and stormy days
We were tossed upon the raging main
And long we strove our bark to save
But all our striving was in vain
E'en then when horror filled my breast
My heart was filled with love for thee
The storm is over and I'm at rest
Oh Mary Mary do not weep
Oh do not weep for me

Oh Mary dear thyself prepare
To meet with me on that bright shore
Where parting hands are known no more
Where we shall meet to part no more
Loud crew the cock the spirit fled
No more of Sandy could she see
But soft the passing spirit said
Sweet Mary Mary do not weep
Oh do not weep for me

Frances Henrietta 1835
Cortes 1847

"Mary's Dream" seems to stand about half way between folk song and parlor song, and the melody is, I think, entirely traditional. The song was very popular all through the nineteenth century.

The *Cortes* and *Frances Henrietta* versions vary just enough from each other and from the standard printed version to indicate that they were being handed on by oral tradition. See *Gems of Scottish Song*, p. 189, and also the *American Musical Miscellany*, pp. 195-198.

THE OCEAN

The ocean hath its silent coves
Deep quiet and alone
Though there be fury on the waves
Beneath them there is none
The awful spirits of the deep
Hold their communion there
And there are those for whom we weep
The young the bright the fair

Calmly the weary seamen rest
Beneath their own blue sea
The ocean solitudes are blessed
For there is purity
The earth has guilt the earth has care
Unquiet are its graves
But peaceful sleep is ever there
Beneath the dark blue waves

Fortune 1840

THOU HAST LEARNED TO LOVE ANOTHER

SONGS THE WHALEMEN SANG

Thou hast learned to love another
Thou hast broken every vow
We have parted from each other
And my heart is lonely now
I have taught my looks to shun thee
When coldly we have met
For another's smile has won thee
And thy voice I must forget
Can I forget thee never
Farewell farewell forever

We have met in scenes of pleasure
We have met in halls of pride
I have seen they new found treasure
I have gazed upon thy bride
I have marked the timid luster
In thy downcast happy eye
I have seen thy gaze upon her
Forgetting I was by
I grieve that e'er I met thee
Fain fain would I forget thee
'Twere folly to regret thee
Farewell farewell forever

We have met and we have parted
But I uttered scarce a word
Like a guilty thing I started
When thy well known voice I heard
Thy looks were cold and altered
And thy words were stern and high
And my traitor courage faltered
When I dared to meet thine eye
Oh woman's love will grieve her
And woman's pride will leave her
Life has fled when love deceived her
Farewell farewell forever

Courier 1842

This very sentimental parlor song seems to have quite a lot of popularity, for there is a very similar version of it in the *Pavilion* journal, 1858.

WE MET 'TWAS IN A CROWD

We met 'twas in a crowd
And I thought he would shun me
He came I could not breathe
For his eyes were upon me
I wore my bridal robe
But I rivaled its whiteness
Bright gems were in my hair
How I hated their brightness

He spoke his words were cold
But his smile was unaltered
I knew how much he felt
For his deep toned voice faltered
He called me by my name
As the bride of another
Oh thou art all the cause
Of this anguish my mother

And once again we met
And a fair girl was with him

He smiled and whispered low
As I once used to hear him
She leaned upon his arm
Once was mine and mine only
I wept for I deserved to feel
Wretched and lonely

And she will be his bride
At the altar he'll give her
The love that was too pure
For a faithless deceiver
And when beside my grave
Your feelings you smother
Forgive as I forgive now
My poor poor mother

Pavilion 1858

Like "Thou Hast Learned to Love Another," this was a parlor song or "song of the boudoir" of the early nineteenth century. But it must have touched home, for all too often a whaleman would return after a voyage of two or three years to find that his intended was indeed the "bride of another."

JAMIE'S ON THE STORMY SEA

PARLOR SONGS THAT WENT TO SEA

Ere the twilight bat was flitting
In the sunset at her knitting
Long a lonely maid was sitting
Underneath her threshold tree
And the daylight died before us
And the vesper air shone o'er us
Fitful rose her tender chorus
Jamie's on the stormy sea

Warmly shone the sunset glowing
Sweetly breathed the young flowers blowing
Earth with beauty overflowing
Seemed the home of love to be
As those angel tones ascending
With the scene and seasons blending
Ever had the same low ending
Jamie's on the stormy sea

Curfew bells remotely ringing
Mingled with that sweet voice singing
And the last red ray seemed clinging
Lingeringly to tower and tree
Nearer as I came and nearer
Finer rose the notes and clearer
Oh 'twas heaven itself to hear her
Jamie's on the stormy sea

Blow ye west wind blandly hover
O'er the bark that bears my lover
Gently blow and bear him over
To his own dear home and me
For when night winds bend the willow
Sleep forsakes my lonely pillow
Thinking on the lonely billow
Jamie's on the stormy sea

How could I but list and linger
To the song and near the singer

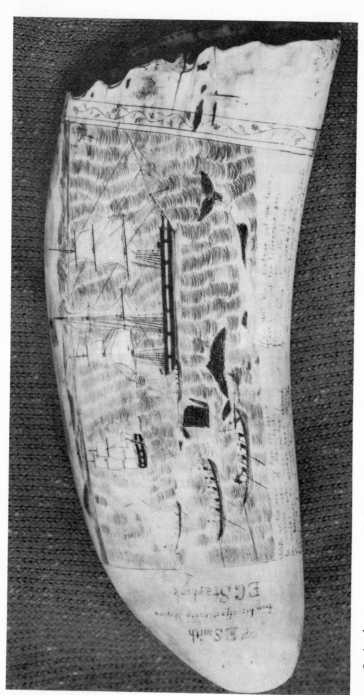

Scrimshaw on a sperm whale's tooth
We worked for our lives
While each tar done his best

Sweetly wooing heaven to bring her
Jamie from the stormy sea
And while yet her lips did name me
Forth I sprang my heart o'ercame me
Grieve no more for I am Jamie
Home returned to love and thee

Euphrasia 1849

This is a literary song pure and simple, and this version shows almost no changes from the original. But because it has a very simple and sweet melody, and because it was in the *Euphrasia* journal, I am including it here as a typical song that the whalemen sang. It can be found in many nineteenth century song collections.

ADIEU TO ERIN

Oh when I breathed a last adieu
To Erin's vales and mountains blue
Where nursed by hope my moments flew
In life's uncoulded spring
Though on the breezy deck reclined

255

SONGS THE WHALEMEN SANG

I listen to the rising wind
What fetters could restrain the mind
That roved on fancy's wind

She bore me to the woodbine bower
Where oft I passed the twilight hour
Where first I felt love's thrilling power
From Mary's beaming eye
Again I watched her flushing breast
Her honeyed lips again were pressed
Again by sweet confession blest
I drank each melting sigh

Dost thou dear Mary my love deplore
And lone on Erin's emerald shore
In memory trace the love I bore
On all our transports dwell
Can I forget the fateful day
That called me from thy arms away
When nought was left me but to say
Farewell my love farewell

Cortes 1847

Perhaps the more common name for this song is "The Emigrant." I haven't been able to find out very much about it. Although the words seem literary it is sometimes hard to tell with Irish songs. The melody is almost surely traditional, and it has a sweet haunting quality.

BLOW HIGH BLOW LOW

Blow high blow low let tempests tear
The mainmast by the board
My heart with thoughts of thee my dear
And love well stored
Shall brave all danger scorn all fear
The roaring wind the raging sea
In hopes on shore to be once more
Safe moored with thee

Aloft while mountains high we go
The whistling winds that scud along
And the surges roaring from below
Shall be my signal to think on thee
Shall be my signal to think on thee
And this shall be my song

And all that night while all the crew
The memory of their former lives
O'er flowing cans of flip renew
And drink their sweethearts and their wives
I'll heave a sigh and think of thee
As the ship rolls through the sea

Cortes 1847

This song was composed by Dibden in 1776 and seems to have been quite popular with seamen as well as in literary circles. Versions quite similar to this may be found in Duncan, pp. 192-193; Kitchiner, pp. 72-73; and Stone, p. 184.

THE ROSE OF ALLENDALE

The moon was fair the sky was clear
Not a breeze came o'er the sea
When Mary left her Highland home
And wandered forth with me
The flowers decked the mountain side
And fragrance filled the vale
By far the sweetest flower there
Was the rose of Allendale

Where e'er I wander east or west
Though fate began to lower
A solace still was she to me
In life's lonely hour
When the tempest lashed our gallant bark
And rent our shivering sail
One maiden form that stood the storm
Was the rose of Allendale

When my fevered lips were parched
On Africk's burning sand
She whispered hopes of happiness
And tales of distant land
My life would been a wilderness
Unblessed by fortune's gale
Had fate not linked my lot with hers
The rose of Allendale

Cortes 1847
Euphrasia 1849
Minerva Smythe 1852

The three versions of this song in the *Cortes, Euphrasia,* and *Minerva Smythe* journals are all pretty similar and all follow the popular parlor version of the song very closely. "The Rose of Allendale" was very

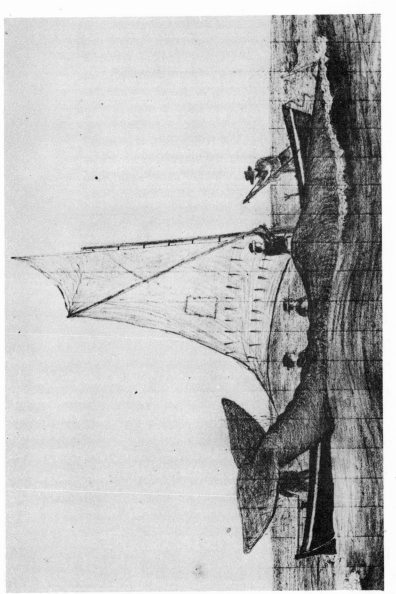

From a journal page
Lay on Captain Bunker
I'm bell for to dart

popular all through the nineteenth century, and will be found in many
song books.

But in the *Nauticon* journal of 1848, there is this little gem, called
"Mary's Cot" which surely is a traditional version of "The Rose of
Allendale."

The morn was clear the morn serene
Not a breath came o'er the sea
When Mary left her Highland cot
To wander forth with me

Come change your ring with me my love
Come change your ring with me
And that will be a token
Whilst I am on the sea

Whilst I am on the sea my love
Not knowing where I am
I'll write to you in letters
From a far and distant land

For I have journeyed o'er many lands
I have sailed on every sea
And Egypt's parching burning sands
No strangers are to me

THE BEACON LIGHT

Darkness was deepening on the seas
And still the hulk drove on
No sail to answer to the breeze
Mast and cordage gone
Gloomy and clear the course of fear
Each looked but for the grave
When full in sight the beacon light
Came streaming o'er the wave

Then wildly rose the gladdening shout
Of all that hardy crew
Boldly they put the helm about
And through the surf they flew
Storms were forgot toil heeded not
And loud the cheer they gave
As full in sight the beacon light
Came streaming o'er the wave

And gaily oft the tale they told
When they were safe on shore
How hearts had sunk and hope grown cold
Amid the bellows roar
That not a star had shone afar
By its pale beam to save
When full in sight the beacon light
Came streaming o'er the wave

Frances Henrietta 1835

Fragments

A FRAGMENT of a song can be a terrible frustration. But those that I have included here, are, I think, too good to be left out of the book.

Some of these songs are fragments because only as much as is given here was included in the logbook or journal. They just stop there. Others are fragments because the rest of the song has been lost with a page or pages torn or cut from the book. And still others are fragments because that is all of the song that the singer could remember.

But there is always the hope that somewhere, somehow, the rest of a song will turn up. Indeed, complete versions of many of these songs do exist, and the references will indicate where they may be found.

A NEW SEA SONG

'Twas in the month of November
It being a stormy day
I shipped on board a Swedish brig
To cross the roaring sea
The captain's name was Ireland
A man both neat and trim
Of officers we had but two
And both was as good as him

FRAGMENTS

The number of the crew was eight
Besides an island cook
And blacker than the common kind
Yet smiling did he look
'Twas on the seventeenth of the month
When we our anchors weigh
And with our topsails sheeted home
For old Barts we bore away

But when eastward clear of Graves
As south south east we steer
When there arose a dismal storm
And for reefing soon we clear
We reefed the topsails fore and aft
And the trysail balance reefed
From head to foot our foresail split
In spite of all our teeth

Herald 1817

Unfortunately this is all there is of "A New Sea Song." No page has been cut from the journal. The song just ends there. And that is too bad because it started out as though it might be very good.

DOWN WAPPING

An honest jack tar oh a-cruising did go
And rolling down Wapping fell in with a beaux
She called him her love and her own true dear
Saying look here my boy can't you raise us some beer

Oh yes my dear girl my own heart's delight
If you'll tell me where we'll find lodgings tonight
Oh yes my dear boy you shall sleep in my arms
For I am the girl to protect you from harm

When these two little lovers away they did gang
Jack taking collections in porter and wine
He rang the bell and asked what's to pay
Ten shillings and six pence the waiter did say

Jack took out his purse and he paid like a man
Says Poll to herself this will answer my plan
For when he's asleep I will lighten his load
So it may not trouble him when on the road

These two little lovers away they did go
Jack merrily whistling gee up and gee wo
It was through a dark alley they chanced for to steer
But Jack was the lad who no colors did fear

Cortes 1847

Unfortunately, this fragment is all that remains of "Down Wapping." For several pages have been cut from the *Cortes* book at this point. Also lost, along with the rest of "Down Wapping" are—according to Histed's index — "The London Butcher," "The Miller's Daughter," "Going Up New York Street," "The Lazy Sailor," and "The Spanish Alphabet." Perhaps some or all of them were bawdy. It was the bawdy songs that were most often cut from the journals.

THE BIBLE STORY

FRAGMENTS

But as she bewailed in sorrowful ditty
The good man beheld and on her took pity
Freemasons are tender so he to the dame
Bestowed an apron to cover her shame

Refrain
Down down down Derry down

Then in process of time men became past enduring
There was nothing but drinking and swearing and whoring
Till Jove being wroth rose up in his anger
And said he would suffer such miscreants no longer

So he from the high windows of heaven did pour
Forty days and forty nights one continuous shower
There was nothing to be seen but waters all round
And in this great deluge most mortals were drowned

Sure never was beheld so dreadful a sight
To see this old world in a very sad plight
See here in the water all animals swimming
Men monkeys priests lawyers cats lap-dogs and women

Here floated a debtor away from his duns
There swam father greybeard naked 'mongst nuns
And here a poor husband quite careless of life
Contented in drowning to get rid of his wife

A thing (?) and a cobler next mingled in view
Of rakes and young spendthrifts there was not a few

SONGS THE WHALEMEN SANG

A whale and a Dutchman came down with the tide
And a reverend old bishop by a young wench's side

For Noah was wisest for Noah judged right
He built up an ark so strong and so tight
Though heaven and earth seemed coming together
He kept safe in his lodge and stood buff to the weather

Then after ye flood like a brother so true
Who still had ye good of ye craft in his view
He delved the ground and planted a vine
He founded a lodge and gave his lodge wine

Then old father Seth he mounted the stage
In manners severe and in masonry sage
He built up two pillars so strong and so thick
The one was of stone and the other of brick

Then change my dear brothers to Blaney's great name
Who's our noble grand master and for virtue famed
That the craft may still flourish and in all quarters spring
Whilst we in full chorus do joyfully sing

Nellie 1769

The opening stanzas of this song must have dealt with creation. They are lost with a page cut from the journal.

Norman Cazden has a version of this song in his *Abelard Folk Song Book,* pp. 109-112, but without the element of Freemasonry. In Barrett, pp. 38-39 there is a song called "The Masonic Hymn" which seems to be another version of "The Bible Story."

Actually this song is probably older than Freemasonry which had its organized beginnings in London about 1717.

FAREWELL MY DEAR NANCY

Your waist is too slender
Your fingers are too small
I fear they would not answer
Our cable ropes to haul
Where the great guns roar like thunder
And the swift bullets do fly
And the silver trumpets sounding
To drown the dreadful cry

If I should meet some fair maids
All blooming fair and gay
If I should take a fancy
What would my Nancy say
What would I say dear William
Why I would love her too
And I would step one side
While she conversed with you

Oh say not so my Nancy
Your words do break my heart
And you and I'll get married
The night before we part
And so these two got married
And he's sailing o'er the main

May heaven's blessings rest on her
Till he returns again

Cortes 1847

The first five stanzas of this song have been lost with two pages cut from Histed's book at this point. Still it is a lovely song and perhaps even this fragment is worth printing although there are complete versions of it in many collections. Two are in Sharp (1), vol. 2, pp. 139-141. See also Sharp (2), pp. 70-71; Mackenzie, pp. 108-112; JFSS, vol. 1, p. 130, vol. 3, pp. 99-100, and vol 7, pp. 50-51. See also JAF, vol. 12, pp. 249-250.

THE TURKEY FACTOR IN FOREIGN PARTS

Behold a ditty 'tis true and no geste
Concerning a young gentleman in the east
Who by much gaming came to poverty
And afterward went many a voyage to sea

Being well educated and with much wit
Three merchants of London did think fit
To make him their captain and factor also
And he on a voyage to Turkey did go

As walking along the streets-o he found
A dead man's carcass lying on the ground
He asked the reason why there he did lie
Then one of the natives made him this reply

This man was a Christian sir when he drew breath
His duties not paid he lies above earth
Then what are the duties the factor cried
Just fifty pounds sir the Turk replied

This is a large sum said the factor indeed
But to see him lie there it makes my heart bleed
So then by the factor the sum it was paid
And under the earth the carcass was laid

From a journal page—Bark *Herald* trying out
  To catch the whales
  And cut and boil

When having gone farther he chanced to spy
A beautiful creature just going to die
A young waiting lady that strangled must be
For nothing but striking a Turkish lady

So thinking of dying with dread she was filled
And rivers of tears like waters distilled
Like a stream or a fountain they did go down
Her rosy cheeks and thence to the ground

Hearing the crime he to end her strife
Said what must I give to save her life
The answer was five hundred pound
Which for her pardon he fairly put down

He said fair creature from weeping refrain
And be of good comfort thou shall't not be slain
Behold I have purchased thy pardon wilt thee
Consent to go to England with me

She said sir you have freed me from death
And I am bound to love you while I breathe
And if you request to England I will go
And my respects to you I will show

He brought her to London and as it was said
He set up housekeeping and she was his maid
There to wait on him and finding her just
With the key to his riches he did her entrust

At last this brave factor was hired once more
To cross the seas where the billows do roar
When into a country he was to steer
Which by her father was governed we hear

Being a (hot) country this maid did prepare
To get fine robes for the country air
She bought a silk waistcoat which I am told
The servant maid flowered with silver and gold.

FRAGMENTS

She said to her master I do understand
You are going factor into such a land
And if that prince's court you do enter in
Be sure this flowered garment's to be seen.

Nellie 1769

Unfortunately the final two or three stanzas of this very nice song are lost with part of a page cut from the journal.

There is a very long version of this song in Flanders and Olney's *Ballads Migrant in New England*, pp. 154-162, much longer than the version in the *Nellie* journal would have been. Probably this version ended on a happy note when the foreign prince recognized the flowered waistcoat as the handiwork of his daughter. See also JAF, vol. 66, pp. 44-45 where it is called "The Turkey Bride."

THE SAILOR BOY'S SONG

Oh I am a Yankee sailor boy
My heart is wild and free
I love a roving sailor's life
As boundless as the sea
I love to watch the vessel as
She dances o'er the tide
I love to see the dolphins play
And frolic by her side
Huzza my heart it leaps with joy
To think I am a sailor boy

Lexington 1853

This is a fragment of a song usually called "Sweet William" or "The Sailor Boy." This song will be found in many collections of folk songs, but there is a particularly nice version of it in Lucy Broadwood's *English County Songs*, pp. 74-75. There are two versions of it in Creighton (1), pp. 89-91 where it is called variously "My Sailor Lad" and "Sailor Bold." See also JAF, vol 31, pp 170-171, and vol. 45, pp. 79-81.

A SAILOR'S TRADE IS A ROVING LIFE

A sailor's trade is a roving life
It's robbed me of my heart's delight
He has gone and left me awhile to mourn
But I can wait till he does return

That short blue jacket he used to wear
His rosy cheeks and his coal black hair
His lips as smooth as the velvet fine
Ten thousand times he has kissed mine

Come father build me a little boat
That o'er the ocean I may float
And every ship that I do pass by
I will enquire for my sailor boy

She had not sailed far o'er (the sea)
Before a king's ship (she did spy)
Captain captain ....
Does my sweet ....

Oh no fair ....
He's ....
On ....
Gives ....

She wrung her hands and tore her hair
Like some female in deep despair
And then her boat to the shore did run
Saying how can I live since my sailor's gone

Come all ye women that dress in white
Come all ye men that take delight
Come haul your colors at half mast high
And help me to weep for my sailor boy

FRAGMENTS

I will sit down and write a song
I will write it both sweet and long
At every line I will drop a tear
At every verse where is my dear

Come dig me a grave both wide and deep
Place a marble stone at my head and feet
And on my breast place a turtle dove
.... know that I died for love

Elizabeth (2) 1847

I am very sorry about the missing lines. They are lost because some of the pages in the journal have had water spilled on them, and where that happened the ink vanished.

HUNTER'S LANE

From Hunter's Lane there came a sound
Which did astonish all around
The dogs did bark the cats did growl
The hogs did squeal and so did fowl

What made the dogs and cats to growl
To think some persons got afoul
In trying to clear the rocks so quick
They raised their voice to the highest pitch

This caused the dogs to madness run
And all the cats they to leave their home
This was the dark and dismal night
That Clarisa made her speedy flight

Euphrasia 1849

I have called this "Hunter's Lane" though it has no title in the journal. Just what it is is hard to say, but it could be the first three stanzas of a song.

FARE YOU WELL

Fare you well you gallant sailors
Since you and I must parted be
If you will prove constant and true hearted
I will prove the same to thee

Chorus
May the winds and waves protect you
To some foreign parts (resign)
And when that you have me my dear don't deceive me
But let your heart be as true as mine

I know my dear you are forced from me
For to cross the raging main
But still my dearest lover
I'm afraid I will never see you again

Galaxy 1827

I have not been able to identify this fragment from the *Galaxy* journal where it is called only "Song." However, it does look like part of a traditional song or ballad.

MOLL BROOKS

I've lost my beau but I care not
He would come back but he dare not
I could have another but I will not
I will be happy and free

Euphrasia 1849

FRAGMENTS

A notation in the *Euphrasia* journal says that this fragment is to be sung to the tune of "Moll Brooks," so I have given it that title. And the melody is "Moll Brooks."

NELSON

Old England expected great news from the fleet
Commanded by Lord Nelson the French to defeat
At length the news came over and o'er England spread
The French they were defeated but Nelson was dead

Cortes 1847

In the *Cortes* manuscript the first six stanzas of this song have been lost with a page cut from the book. That is bad, but this is worse: in JFSS, vol 3, p 273 there is the same final stanza as Histed's and that is all. There the other stanzas of the song are lost because the collector did not consider them worth printing. I can only hope and trust that they are preserved somewhere in Cecil Sharp house in London.

THE BONNET OF BLUE

A ship's crew of sailors from London did steer
On a voyage to the West Indies they were bound
There is one lad among them that I wish I never knew
He is my bonny sailor lad with his jacket so blue

Very early one morning as I arose from my bed
I called for my Betsy my chief serving maid
Saying come now and dress me nice as your two hands can do
And I'll run away with my sailor lad with his jacket so blue

Now as I walked down all by the sea side
My love like the lightning away she did glide
And I for my true love oh what could I do
And away went my heart with her bonnet so blue

She says honored sailor I will be your discharge
I'll free you from a man-of-war and set you at large
And if you'll prove constant and always prove true
You'll never set a stain on my bonnet so blue

But I have a fair girl in my own country
And I never could forsake her for her poverty
For she always proved constant and always proved true
I would never set a stain on her bonnet so blue

Catalpa 1856

This song in the *Catalpa* journal is garbled and full of mistakes
so that I have had to supply two or three words even to get what we

have here. In the *Galaxy* journal (1827) there is a two-stanza fragment of this same song as follows:

> I have a dear lass in my own country
> I never will slight her for her poverty
> Here's to the girl I love I will always prove true
> I will never put a stain on my bonnet so blue

> I send for him from London to Hull
> I have my love's picture drawn out in full
> To hang in my bed chamber often to view
> To think of my boy and his bonnet so blue

There is a very nice version of this song in Robert Ford's *Vagabond Songs and Ballads of Scotland,* vol. 2, pp. 1-3 for those who want the full story. See also Kidson (1), pp 118-119 where it is called the "Bonny Scotch Lad." Also JFSS (1931), p. 191.

WHEN I REMEMBER

> When I remember all
> The girls I've met together
> I feel like a rooster in the fall
> Exposed to every weather
> I feel like one who treads alone
> Some barnyard all deserted
> Whose oats are fled
> Whose hens are dead
> Or off to market started

Ocean Rover 1859

COME LET US BE JOLLY

> A sailor sighs as sinks his native shore
> And all the lessening turrets' beauty fade
> He climbs the mast to feast his eyes once more
> And busy fancy fondly lends her aid

But grieving is a folly
Come let us be jolly
We've troubles at the sea boys
We've pleasures on shore

Argus 1813

This has no title in the *Argus* journal. However, it does look as though it may be part of a song, with the second stanza acting as a chorus.

YE PARLIAMENTS OF ENGLAND

Ye Parliaments of England
Ye lords and commons too
Consider well what you're about
And what you mean to do

You're now at war with Yankees
I'm sure you'll rue the day
You roused the sons of liberty
In North America

FRAGMENTS

You first confined our commerce
And say our ships can't trade
You next confined our seamen
And used them as your slaves

You then insulted Rogers
While cruising on the main
And had we not declared war
You'd done it over again

You thought our frigates were but few
And Yankees would not fight
Until brave Hull your Gourier took
And banished from your sight

The Wasp then took the Frolic
You nothing said to that
The Poitiers being of the line
Of course she took her back

Trident 1846

This is a fragment of one of the most famous of all Yankee sea songs. For more complete versions see Colcord, pp. 128-130, and Flanders, pp. 195-196.

OLD HORSE

Old horse old horse what brought you here
You've carted stone for many a year

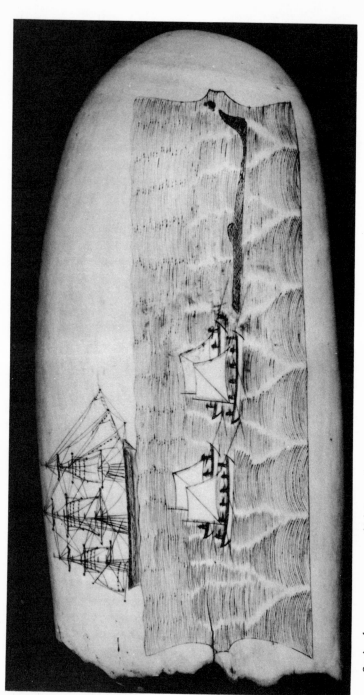

Scrimshaw on a sperm whale's tooth
We'll tow him alongside
And rob him of his hide

FRAGMENTS

Till killed with blows and sore abuse
They salt you down for sailor's use

The sailors they do me despise
They turn me over and damn my eyes
Cut off my meat and pick my bones
And pitch the rest to David Jones

Carthage 1842

This fragment deals with the seaman's perpetual gripe against the food on shipboard. The salt meat was usually called salt horse, and how often it actually was salt horse is a good question.

Last But Not Least

I THINK that some of the best songs in the book are in this section, which includes a little bit of everything. Some might have been included in other categories, but for one reason or another it seemed best to group them together here.

A few are very hard to classify. Others are easy as, white spirituals, gospel songs, music hall songs, and so on. Two or three may not even be songs at all, just verse, but I don't think so. As one goes through the logbooks and journals, one somehow seems to sense when a song is a song and when it isn't.

AN ANCIENT RIDDLE

Adam God made out of dust
But thought it best to make me fust

LAST BUT NOT LEAST

So I was made before the man
To answer God's most holy plan

My body God did make complete
But without arms or legs or feet
My ways and acts he did control
But to my body gave no soul

A living being I became
And Adam gave to me my name
I from his presence then withdrew
And more of Adam never knew

I did my master's law obey
Nor from it never went astray
Thousands of miles I go in fear
But seldom on the earth appear

For purpose wise which God did see
He put a living soul in me
A soul from me my God did claim
And took from me that soul again

For when from me the soul had fled
I was the same as when first made
And without hands or feet or soul
I travel on from pole to pole

I labor hard by day and night
To fallen man I give great light
Thousands of people young and old
Will by my death great light behold

No right nor wrong can I conceive
The scriptures I can not believe
Although my name therein is found
They are to me an empty sound

SONGS THE WHALEMEN SANG

No fear of death doth trouble me
Real happiness I ne'er shall see
To heaven I shall never go
Or to the grave or hell below

Now when these lines you slowly read
Go search your Bible with all speed
For that my name's recorded there
I honestly to you declare

Smyrna 1853

"An Ancient Riddle" has this preface in the journal: "A great many years ago a prominent merchant of Taunton promised to an eccentric old woman named Lucy King, living in the neighboring town of Berkley a desirable prize if taking her subject from the Bible she would compose a riddle which he could not guess. She won the prize with the following."

I had not thought that this was truly a song, but merely some better than average original verse. But I was wrong, for it is indeed a song. In *The Journal of American Folklore*, vol. 49, pp. 240-241 there is a version of "An Ancient Riddle" collected by Bess Alice Owens from the singing of Miss Sally Howard of Lynn Grove, Kentucky, and it is the same song as the one in the *Smyrna* journal. Miss Sally Howard's melody, which I am using here, is simple and has something of the feeling of a white spiritual.

The answer to the riddle does not seem to have been collected by Miss Owens, but it is in the *Smyrna* journal as follows:

God made the whale without a soul
To wander wide from pole to pole
But when the fish had Jonah swallowed
As in the deep the prophet wallowed
It had a soul enshrined within
Till prayer received relief from sin
The prophet sent to save the dead
As light thus shone on God's command
So whales shed light o'er all the world

LAST BUT NOT LEAST

To all mankind the slave and queen
Unless they burn the kerosene.

And kerosene was a very bad word to whalemen. Naturally.

WAIT FOR THE WAGON

Will you come with me my Phyllis dear
To yon blue mountain free
Where the blossoms smell the sweetest
Come rove along with me
It's every Sunday morning when I am by your side
We'll jump into the wagon and we'll all take a ride

Chorus

Wait for the wagon wait for the wagon
Wait for the wagon and we'll all take a ride

285

Where the river runs like silver
And the birds they sing so sweet
I have a cabin Phyllis dear
And something good to eat
Come listen to my story it will relieve my heart
So jump into the wagon and off we will start

Do you believe my Phyllis dear
That rich old man with all his wealth
Can make you half so happy dear
As I with youth and health
We'll have a little farm there a horse a pig a cow
And you will mind the dairy while I guide the plow

Your lips are red as poppies
Your hair so sleek and neat
All braided up with dahlias
And hollyhocks so neat
It's every Sunday morning when I am by your side
We'll jump into the wagon and all take a ride

Together on life's journey
We'll travel till we stop
And if we have no trouble
We'll reach the happy top
So come along my Phyllis dear and be my happy bride
We'll jump into the wagon and all take a ride

Smyrna 1853

"Wait for the Wagon" will be found in many old song books, but today it is much better known as a square dance and fiddle tune than as a song.

PRAYER

To thee of God whose awful voice
Earth sea and air obey

LAST BUT NOT LEAST

This humble hour of prayer we raise
And here our homage pay

We've seen thy works upon the sea
Thy wonders in the deep
When thou didst loose the stormy winds
O'er raging waves to sweep

We've sunk in ocean's fearful depths
Then rose on mountain wave
We've hung on brink of dread abyss
That yawned like watery graves

When from the deep we called on God
The raging winds to stay
The raging winds were hushed to sleep
At his almighty sway

Our perils o'er our ship at rest
With joy we tread these courts
Here let us raise our songs of praise
Where God himself resorts

And when again we spread our sails
And o'er the billows roll
We'll bear the gospel's joyful light
Till seen from pole to pole

Uncas 1834

This is probably original. But because it is a good prayer and because it shows one side of many whalemen's natures that is often too little recognized, it is included here.

Whether or not there was any recognized religious service on a whaler seems to have depended almost entirely on the piety of the skipper. On some vessels there were regular Sunday morning services with Bible reading and prayer in between times. And there were even some captains who would not permit a boat to be lowered on Sunday. But by and large that was considered carrying it a little too far.

287

The few gospel songs that follow are a good sample of those that are found in the journals.

THE PILOT

Oh pilot 'tis a fearful night
There's danger on the deep
I'll come and pace the deck with thee
I do not dare to sleep
Go down go down the pilot cries
This is no place for thee
Fear not but trust in Providence
Wherever thou mayst be

Oh pilot dangers often met
We all are apt to slight
And thou hast known the raging seas
But to subdue their might
Oh 'tis not apathy he cried
That gives this strength to me

LAST BUT NOT LEAST

Fear not but trust in Providence
Wherever thou mayst be

On such a night the sea engulfed
My father's lifeless form
My only brother's bark went down
In just so wild a storm
And such perhaps may be my fate
But still I say to thee
Fear not but trust in Providence
Wherever thou mayst be

Lotos 1833
Walter Scott 1844

THE LORD OUR GOD

The Lord our God is full of might
The winds obey his will
He speaks and in his heavenly height
The rolling sun stands still

Rebel ye waves and o'er the land
With threatening aspect roar
The Lord uplifts his awful hand
And chains you to the shore

Howl ye winds of night your force combine
Without his high bequest
Ye shall not in the mountain pine
Disturb the sparrow's rest

Ye nations bend in reverence bend
Ye monarchs note his nod
And bid the jovial song ascend
To celebrate our God

Addison 1834

ROW ON

Row on row on another day
May shine with brighter light
Ply ply the oars and pull away
Thou must not come tonight

Clouds are upon the summer sky
There's thunder on the wind
Pull on pull on and homeward hie
Nor give one look behind

Bear where thou goest the words of love
Say all that words can say
Changeless affections strength to prove
But speed upon the way

Oh like yon river would I glide
To where my heart would be
My bark should soon outsail the tide
That hurries to the sea

But yet a star shines constant still
Through yonder cloudy sky
And hope as bright my bosom stills
From faith that can not die

Row on row on God speed the way
Thou must not linger here
Storms hang about the closing day
Tomorrow may be clear

Three Brothers 1846

So far I have not been able to find this song in print. In Duncan's the *Minstrelsy of England* there is a song called "Jog On, Jog On" the opening line of which reads, "Jog on, jog on the footpath way." But very probably there is no connection between the two songs.

THE RECRUITING SARGEANT

From Paphos' gentle fold I come
To raise recruits with merry fife and drum
The queen of beauty here by me invites
Each nymph and swain to taste sweet delights
Obey the call and seek the happy land
Where captain is Cupid and bears sole command

Ye nymphs and ye swains who are youthful and gay
Attend to my song and be blest while ye may
Come lady and lass to the sound of the drum
I have treasure in store which you never have seen
Then soft let us rove to the garland of love
Where Cupid is captain and Venus is queen

Each nymph of sixteen who would fain be a wife
Shall soon have a husband to prosper her life
The lasses come hither to the sound of the drum
I have sweethearts in store which you never have seen
Then haste let us rove to the garland of love
Where captain is Cupid and Venus is queen

Would a swain be blest with a nymph to his mind
Let him enter with me and his wish he shall find
I can bless him for life with a kind loving wife
More beautiful far than nymph ever seen
Then haste let us rove to the garland of love
Where captain is Cupid and Venus is queen

In Paphos we know no discord nor strife
And each nymph and each swain may be happy for life
And transports of joy we each moment employ
And taste such delights as were never yet seen
Then haste let us rove to the garland of love
Where Cupid is captain and Venus is queen

Nellie 1769

SONGS THE WHALEMEN SANG

The handwriting here is very careless and there are blots on the page to make it worse, so I cannot be sure of some of the words. "Rove," the fifth word in the next to the last line of all the stanzas save the first, is particularly tentative. Paphos was a city in Cyprus sacred to Venus.

THE POST BELOW

What means this rout this noise this roar
Have you ne'er seen a rogue before
Are villains then indeed so rare
Ye must needs press and gape and stare
Come forward ye who look so fine
Wily gains is illy got as mine
Step up you'll soon reverse the show
The crowd above the few below

Well for my knavery here I stand
A spectacle to all the land
High elevated on the stage
The greatest rascal of this age
And for the mischief I have done
Must put this wooden neck-cloth on

There how his heaving back is stripping
Quite callous grown from frequent whipping
In vain ye'll wear your whip cord out
Ye'll n'er reform a rogue so stout
To make him honest take my word
You must apply a bigger cord

Now all that see this shameful sight
That ye may get some profit by't
Keep constantly in mind I pray
The few words that I have to say
Follow my steps and you may be
In time perhaps advanced like me

LAST BUT NOT LEAST

Or like my fellow laborer trow
May get perhaps a post below

Frances Henrietta 1835

This strange song has no title in the *Frances Henrietta* journal. It must be old for the stocks and whipping post have not been used in New England for a long time. Two whole lines seem to be lost from both the second and third stanzas.

A LOVE SONG IN THE YEAR 1769

Down by a shady bower nigh to a pleasant green
Where flags and flowers adorned most pleasant to be seen
I heard sad lamentation a poor disturbed swain
Who said his love sick passion doth torture him with pain

I note what a great vexation that she shall frown on me
Could I but speak of that passion how happy I would be
Could I but gain her favor her pretty lips to kiss
I should be crowned forever with joy and happiness.

Oh if you will but marry me you may as well me tell
And if you will not have me there is some other will
So farewell pretty Nabby till we do meet again
It runs all in my fancy to cross the raging main

You fairest of all creatures how can you cruel be
The birds are of good nature that fly from tree to tree
More kinder to each other how can you prove so coy
To make a wounded lover and all his hopes destroy

Cupid be more kinder and cure me of my pain
It ne'er was in my nature a woman to disdain
I can not blame those we love although they prove so coy
For a man is made to crave and a woman to deny

Nellie 1769

293

From a journal page—three whalers gamming
And if you should gam her
Just bear it in mind

LAST BUT NOT LEAST

The name of this song is the one Peter Pease gives it in the *Nellie* journal. Although I have looked high and low for it I have not been able to find any other sure version.

ELEGY ON THE DEATH OF A MAD DOG

Good people all of every sort
Give ear unto my song
And if you find it wondrous short
It can not hold you long

In Ixlington there was a man
Of whom the world might say
That still he was a Godly man
When e'er he went to pray

A kind and gentle heart he had
To comfort friends and foes
The naked every day he clad
When he put on his clothes

And in that town a dog was found
As many a dog there be
Both mongrel puppy whelp and hound
Of high and low degree

The dog and man at first were friends
But when a pique began
The dog to gain some private end
Went mad and bit the man

Around from all the neighboring streets
The wondering neighbors ran
They swore the dog had lost his wits
To bite so good a man

The wound it seemed both sore and sad
To every Christian's eye
And while they swore the dog was mad
They swore the man would die

But soon a wonder came to light
That showed the rogues they lied
The man recovered of his life
It was the dog that died

Diana 1819

In the *Laurel Song Book* published in Boston in 1901 this song is credited to Oliver Goldsmith.

A CHARMING FELLOW

Lord what care I for mom or dad
Why let them scold and bellow
For while I live I'll love my lad
He's such a charming fellow

The last fair day on yonder green
The youth he danced so well-o
So spruce a lad was never seen
As my sweet charming fellow

The fair was o'er night was come on
The lad was somewhat mellow

LAST BUT NOT LEAST

Says he my dear I'll see you home
I thanked the charming fellow

We trudged along the moon shone bright
Says he if you'll not tell-o
I'll kiss you here by this good light
Lord what a charming fellow

You rogue says I you've stopped my breath
Ye bells ring out my knell-o
Again I'd die so sweet a death
With such a charming fellow

Ann 1776

THE WREATH

There went a maid and plucked the flowers
That grew upon a sunny lea
A lady from the greenwood came
Most beautiful to see

She met the maiden with a smile
And twined a wreath into her hair
It blooms not yet but it will bloom
So wear it ever there

And as the maiden grew and roamed
Beneath the moon so pale and wan
And tears fell from her sad and sweet
The wreath to bud began (?)

And when a joyous bride she lay
Upon her faithful leman's breast
Then smiling blossoms burst the folds
Of their encircling nest

Soon cradled gently in her lap
The mother held a blooming child

And many a golden fruit from out
The leafy chaplet smiled

And when alack her love had sunk
Into the dark and dusky grave
In her disheveled hair a sere
Dry leaf was seen to wave

Soon she too beside him lay
But her beloved wreath she wore
And it oh wondrous sight to see
Both fruit and blossoms bore

Cortes 1847

I CAN NOT CALL HER MOTHER

The wedding rite is over
And though I turn aside
To keep the guests from seeing
My tears I can not hide

I wreathed my face in smiling
As I led my little brother
To greet my father's chosen one
But I could not call her mother

My father in the sunshine
Of brighter days to come

298

LAST BUT NOT LEAST

May half forget the shadow
That darkens our old home

His heart is no more lonely
But I and little brother
It's orphans we shall ever be
God can give us but one mother

They have born my mother's picture
From its accustomed place
And placed beside my father's
A younger fairer face

And they have made her chamber dear
The boudoir of another
But I can not forget her
My own my angel mother

Lexington 1853

This very sad song seems to have had wide currency. Perhaps even wider than the printed versions of it would indicate, for it is not the type of song that folklorists generally admire. It savors too much of the music hall and carnival singers.

There is an even sadder version of the song in Kincaid (2), p. 32, where it is called "The Blind Girl."

THE VILLAGE BORN BEAUTY

See that star-breasted villain to yonder cot bound
Where the sweet honeysuckle entwineth around
But sweeter far sweeter than flowers ever seen
Is the poor hedger's daughter the pride of the green

Never more never more will she there please all eyes
Her peace of mind withers her happiness dies
She (franseth) and trembles and yet dares to roam
The village born beauty is sundered from her home

SONGS THE WHALEMEN SANG

With her fast chaise and four now in London sits down
She is robbed of all virtue and launched on the town
With her carriage and servants and friends so gay
The village born beauty she now bears the (sway)

To the opera the playhouse the park or elsewhere
This beauty outrivals all beauty that's there
I begged her for mercy her downfall to tell
The village born beauty she yet reigns the belle

Through streets lanes and alleys she now takes her way
Exposed to the weather by night and by day
Cold houseless and shivering and wet to the skin
While glass after glass drowns her sorrow in sin

Take pity ye kind ones and do not be severe
Pray give to your sister the boon of a tear
It's hard to condemn her so leave her a while
For the heart often bleeds when the face wears a smile

In some lone garret or work house obscure
On a bed made of rags spread out on the floor
She sends up her prayers a poor sinner to save
The village born beauty is laid in her grave

Ocean Rover 1859

AS I GROW OLD

LAST BUT NOT LEAST

If I live to grow old as I find I go down
Let this be my fate in a fair country town
May I have a warm house with a stone at my gate
And a cleanly young girl to rub my bald pate

Chorus

May I govern my passions with an absolute sway
And grow wiser and better as my strength wears away
Without gout or stone by a gentle decay

In a country town by a murmuring brook
With the ocean at a distance on which I may look
With a spacious plain without hedge or stile
And an easy passage to ride out a mile

With Horace and Plutarch and one or two more
Of the best of poets lived in ages before
With a dish of roast mutton not venison or lamb
And clean tho course linen at every meal

With a pudding on Sunday with stout humming liquor
And a remnant of Latin to puzzle the vicar
With a hidden reserve of Burgundy wine
ᵀo drink the President's health as oft as we dine

With courage undaunted may I face my last day
And when I am dead may the better sort say
In the morning when sober in the evening when mellow
He's gone and hain't left behind his fellow

Final Chorus
For he governed his passions with an absolute sway
And grew wiser and better as strength passed away
Without gout or stone by a gentle decay

Paulina 1808

In *Calliope,* London, 1788, p. 58, this song is called "If I Live to Grow Old." And there it is the King's health and not the President's which is toasted in Burgundy wine.

NORAH DARLING

I am going far away Norah darling
And I'm leaving such an angel far behind
It will break my heart in two
Which I fondly gave to you
For there's no one else so loving kind and true

So come to my arms Norah darling
Bid your friends in dear old Ireland good bye

LAST BUT NOT LEAST

And it's happy you will be
In that dear land of the free
Living happy with your Barney McCoy

I would go with you Barney darling
But the reason why I've told you oft before
It would break my mother's heart
If from her I had to part
And go roaming off with you Barney McCoy

I am going far away Norah darling
Just as sure as there's a God that I adore
And remember what I say
That until the judgement day
You will never see your Barney any more

I would go with you Barney darling
If my mother and the rest of them were there
For I know we would be blest
In that dear land in the west
Living happy with my Barney McCoy

I am going far away Norah darling
And my ship it now is anchored in the bay
And before tomorrow's sun
You will hear the signal gun
So be ready it will carry us away

So come to my arms Norah darling
Bid your friends in dear old Ireland good bye
For it's happy you will be
In that dear land of the free
Living happy with your Barney McCoy

A Hicks 1879

I have very little information on this song except that it is some-times called "Barney McCoy." It was printed as sheet music in the mid nineteenth century, and I doubt if it can be much older than that.

SONGS THE WHALEMEN SANG

THE SAILOR'S FAREWELL

List ye winds while I repeat
One parting signal to this fleet
One farewell ere I roam

Farewell to father reverend hulk
In spite of guns in spite of bulk
Must soon his cable slip
But ere he lies quite broken by
I'll task my mind and this I'll try
The flag of gratitude to fly
And duty to this ship

Farewell to mother first rate she
Who launched me on life's stormy sea
And rigged me fore and aft
May Providence her timbers spare
And keep her in a good repair
To guide the younger craft

Farewell to Sarah lovely she
And may her voyage on life's troubled sea
Serene and tranquil prove
And as she the bonds of wedlock wears
May no rude hand asunder tear
Those ties that heaven ordained

Farewell to Ann lovely barge
With figurehead both high and large
And fitted out so fine
If taken by a sailor bold
She'll lead the van and all control
A first rate of the line

Farewell to Virginia lovely yacht
But whether she'll be manned or not
I can not now foresee

LAST BUT NOT LEAST

May some good ship her tender prove
Well fitted with stores of truth and love
And take her under lee

Farewell to William the privateer
With narrow hull and classic gear
And collier rigged and classed
If for a wife abroad he roams
Some rich galleon he'll run aground
And make his grappling fast

Farewell to John the slender smack
Though now he's on a different tack
In docks near laid aside
Like me from home he's now adrift
And without ballast learns to shift
And stem against the tide

Farewell to Samuel the jolly boat
And the little ones at home afloat
When they arrive at sailing age
May wisdom be their weather gauge
And guide them on their way

Farewell once more
A rude rough tar
Requests the parting tribute of a tear
And should my bark this voyage be lost
He hopes in heaven to meet at last

Lotos 1833

This version of the song from the *Lotos* journal is pretty badly disarranged and garbled. However it seems to be truly traditional. Here is a shorter version of it from the journal of the ship *Elbe* of Poughkeepsie on a whaling voyage to the Pacific in the years 1833-1837. This is called "Thomas, Farewell."

Await ye winds till I repeat
One parting signal to the fleet

One farewell ere I roam
For where shall I a convoy find
Like those dear ships I've left behind
When out adrift from home

Farewell to father reverend hulk
Who spite of force in spite of bulk
Must soon his cable slip
But ere he's broken up I'll try
The flag of gratitude to fly
In duty to the ship

Farewell to mother first rate she
Who launched me on life's stormy sea
And rigged me fore and aft
May Providence her timbers spare
And keep her hull in good repair
To tow the smaller craft

Farewell to John the Privateer
Narrow built with classic gear
And collier rigged and classed
When for a wife he cruises round
Some rich prize he will run aground
And make his grapline fast

Farewell to Sarah lovely yacht
But whether she will be manned or not
I cannot now foresee
May some good ship a tender prove
Well found in stores of truth and love
And take her under his lee

TERRIBLE POLLY

LAST BUT NOT LEAST

A song a song to everyone
I mean to write it down with speed
All you that have not your wits learned
For shame you stand in need

It is I that never made songs before
And I mean to make this and never make more
And it is woman wicked woman what will this world come to
Oh Polly terrible Polly what do you mean to do

The girls have got so far from (truth)
'Tis after them their mother
They will sooner bilder that bonnets on their heads
'Tis one I love another

They will go to church and make such a show
I can't see my God for looking at you
'Tis woman woman wicked woman what do you want to do
Oh Polly Polly terrible Polly what will this world come to

They will patch they will paint they will powder their locks
And let their titties go bare
That the young men may plainly see
That they be maidens fair

They will patch they will paint and powder besides
Methinks the devil is in woman for pride
Woman woman wicked woman what will this world come to
Polly Polly terrible Polly what do you mean to do

If ever I marry whilst I live
I will have some of the oldest fashion

And she shall be as good as she looks
I will have none of their new transformation

They will fix their hats and caps so nice
And in them they carry a bucket of lice
Oh it's woman woman wicked woman what do you mean to do
Polly Polly terrible Polly what will this world come to

For women wear jackets much like unto men
And they will wear britches when they can get them
And it is woman woman wicked woman what will this world come to
Oh Polly Polly terrible Polly what do you mean to do

Herald 1817

This song does not have a title in the *Herald* journal, and I have not found any version of it in print that I can surely identify. Speculation can be very dangerous where folk songs are concerned, but I just wonder if it may not be a much altered variant of the white spiritual "Wicked Polly" in Lomax' *American Ballads and Folk Songs*.

THE WIDE WORLD OF WATERS

From the wide world of waters
In rapture I came
To the land of my fathers
My own native home
O'er the billows I've wandered
And fain could I say
How fondly I've pondered
On home when away

The anguish that's burning
Oh what can remove
Like the thoughts of returning
To those whom we love
When our seamen were sailing
How drear o'er the main

LAST BUT NOT LEAST

In my hammock oft dreaming
Of home have I lain

At my own native cottage
The wild woods along
In fancy I've listened
To hear the birds song
But all the fond revries
That flit through my brain
Were precoursers of sorrow
Forerunners of pain

For there stands my cottage
With brambles grown wild
I seek for my partner
I ask for my child
And ye point to their graves
Oh where shall I flee

....
....

Bengal 1832

This song has no title in the journal, and I have not yet been able to find it in print. Ira Poland's handwriting is extremely difficult at times and I could not decipher the last two lines.

SONG OF SOLOMON'S TEMPLE

In scripture we read of an extraordinary old king
The monarch of Israel his preaises we will sing

Scrimshaw on a sperm whale's tooth
Now our boats being lowered there arose a contest
Among the boats crews t' see which should do best

LAST BUT NOT LEAST

He built up a fine fabric as we understand
It's way on Mount Moriah called Jerusalem

It may him that slew Goliath in scripture we find
He mustard all the workmen to accomplish his desire
He ordered King Solomon on being his son
To finish the building that he had begun

King Solomon was ordered to execute his plan
He numbered all the workmen that was in the land
Four thousand five hundred barbarians he kept in reserve
Eight thousand in the mountains to cut him and carve

Four thousand five hundred was chosen to be
Master of workmen to own ... (?) ... see
And if you believe me it is certainly true
For he clothed them all in the orange and blue

It was the cunning workmen their stones they did square
All fit for the building before they came there
And on proper carriages they carried them down
That on the fine building no hammer should sound

They built up two chariots of the emage whise (?)
They stretch forth their wings to carry the ark
The length of their wings reached King Solomon's porch
So he might behold them as he went to church

And when the queen of Sheba she heard of his fame
Straight on to Jerusalem she instantly came
And when she came there she was struck with surprise
The sight of the building it dazzled her eyes

She asked him questions according to art
He answered them all that belonged to his part
For wit and for learning none could him excel
And so the queen of Sheba she loved him well

Here's a health to all freemasons that dwell in love
And to those that hold the grand lodge above
Here is a health to king Hiram and king Solomon also
Come fill up the bumper and we will drink and huzza

Galaxy 1827

Mackenzie, pp. 381-383 states that this song is rather rare. There
it is called "The Building of Soloman's Temple." See also a version of the
song in *Howe's* 100 *Old Favorite Songs*, p. 262.

POLL AND SAL

When I was young and in my prime
A-courting I was much inclined
I lived along with my master
Till I grew up a long feller

Refrain

Diddle I dum diddle I dee

Sunday night the moon shined bright
All the stars gave forth their light
As I was advancing over the wing
I thought I heard a fair maid sing

LAST BUT NOT LEAST

A windy night I sat out again
I met a fair maid on the plain
Then as she sees I had no nag
She instantly gave me the bag

The next night I sat out with whip and spur
A-going on purpose to cheat her
And when she thought I rode that night
She let me stay till broad daylight

When I got home 'twas just about nine
Just about at coffee time
My master he looked very cross
Because he thought I rode his horse

Wednesday night I began to nod
Wishing for some place to lodge
As I sat a-thinking by the fire
I tumbled back within my chair

The first I hit was my head on the floor
Which made me so I slept no more
My mistress she stood side of me
My mistress she laughed heartily

Thursday night I sat out again
I went a-visiting cousin John
And we both went to see the girls
He took Poll and I took Sal

In the night the war broke out
The old woman came sidling out
Then she took both of us by our hairs
And we come tumbling down the stairs

She says begone you bougey boys
For I will have none of your ways

I set too much store by both my gals
To let you stay with Poll or Sal

I looked east and I looked west
To see which of them I liked the best
But we both getting turned out the door
Was worse than getting the bag before

As we were going up to a door
I heard someone say my face I will scour
For I am not fit to be seen
And then she stepped behind the screen

Some was relating of their yarn
And some their stockings they did darn
Some took snuff and some did not
How many there was I have forgot

Some was up and some was abed
Some was under the coverlet
Sometimes I stayed sometimes I would not
Sometimes I might sometimes I could not

Then I set out in good earnest
I courted a daughter of the priest
And then a bargain soon was made
And we were quickly married

Herald 1817

This song of adolescent courting must be an early version of the song that is usually called "When the Boys Go A-Courting." The element of trying to cheat the girls is in both songs as is also the idea that the girls are slovenly and unkempt. Here the coffee break comes a little early for modern living.

For versions of "When the Boys Go A-Courting" see Downs and Seigmeister, pp. 182-183; Lomax and Lomax, *Best Loved American Folk Songs*, p. 42; and Sharp (1), vol. 2, pp. 205-206. Also JAF, vol. 31, where the song is called "Over the Mountains."

THE KEYHOLE IN THE DOOR

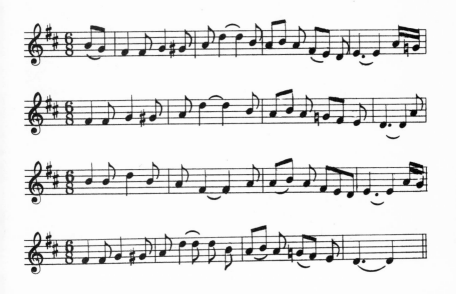

We left the parlor early
I think it was scarce nine
When by the chance of fortune
Her room was next to mine
Resolved like bold Columbus
New regions for to explore
I took a strange position
By the keyhole in the door

In bending down in silence
And resting on my knee
Most patiently I waited
To see what I could see
She first took off her collar
It rolled upon the floor
And I watched her stoop to get it
Through the keyhole in the door

SONGS THE WHALEMEN SANG

Sweet Jenny then proceeded
In taking off her dress
And most of her under garments
Some fifty more or less
But to speak the truth sincerely
I think there was a score
But I could not count exactly
Through the keyhole in the door

She then unloosed her tresses
Her waving chestnut hair
Which fell in streaming torrents
All down her shoulders bare
Then quickly she rebound them
More firmly than before
While I watched this witching process
Through the keyhole in the door

Then down upon the carpet
She sat with graceful ease
And raised her spotless linen
Above her snowy knees
Two dainty sky-blue garters
On either leg she wore
And they made a charming picture
Through the keyhole in the door

She then approached the fire
Her dainty limbs to warm
And nothing but her shimmey
Concealed her lovely form
Thinks I take off that shimmey
I ask for nothing more
Ye gods I saw her do it
Through the keyhole in the door

And then with nimble fingers she
Donned her snow white gown

LAST BUT NOT LEAST

And on her bed sweet Jenny
Prepared to lie her down
Thinks I a bed so ample
Might hold at least one more
But I did not dare to say it
Through the keyhole in the door

Then down upon the pillow
She laid her lovely head
The light she then extinguished
And darkness veiled the bed
No use in waiting longer
I knew the show was o'er
So my post I then abandoned
By the keyhole in the door

So come ye men of science
Why strain your eager eyes
A-gazing at the planets
That alumernates the skys
For there are greater wonders
Than you know of in your lore
For a telescope is nothing
To a key hole in the door

A Hicks 1879

Although this song has been pretty widely known and sung, I have never seen it in print. The reason must be that it is not what one would call in the best of taste. The melody here is the one I have always heard it sung to.

THE SANDSHARK

The sandshark feeds on the sailor bold
Till his throbbing heart grows poor and old
Or swallows him down while he's tender and young
And laps his blood with a greedy tongue

317

SONGS THE WHALEMEN SANG

Not where the upland fountains play
Not where the timid minnows stay
But close by the surf of the mighty deep
In the gulf of hell does the sandshark sweep

By the devil's reef at noon and night
In the alley dark and the bar room bright
Quick as the victim comes to the lair
He is clutched by the gory monster there

What cares he for the sailor's cries
For the father's groans or the mother's sighs
He has sought that sailor from pole to pole
And the sandshark eats him body and soul

In the sandshark's haunts there is music heard
And the sparkling waves of the bowl are stirred
And the siren's lust and the gambler's spell
Are thick by the sandshark's road to hell

And seldom shall ever the victim pass
From the harlot's grip and the demon glass
For at home and abroad through every flood
The sandshark waits for the sailor's blood

Pavilion 1858

This is a strange song indeed, and I have not yet found it in print. Ken Goldstein thinks that it may be an English music hall song.

Actually the sandshark is a harmless beast, but the dangers on shore that faced the poor sailor from shark-like humans were very real.

BILLY O'ROURKE

Faith I cut my sticks and brushed my brogues
At the latter end of May sirs
And off to Dublin town did trip
To sail upon the seas sirs
To see if I could get employ

LAST BUT NOT LEAST

To cut their hay or corn sirs
To pick up pence upon the seas
The cockneys I might learn sirs

Refrain

Philliloo and etc.

I gave the captain six thirteens
To carry me over to Porgate
But before I got one half the way
It blew to a devilish hard rate
Says the captain says he to the bottom you'll go
Says I I don't care a damn farthing
I payed you to carry me to Porgate you know
And I'll make you stick to your bargain

Oh the great big stick that grew out of the ship
It began to roar and whistle
And one and all both great and small
Cried Paddy you'll go to the devil
I put a girl upon my back
And I jumped into the water
Oh murder Pat what are you at
But safe to land I brought her

I met an honest gentleman
A travelling the road sirs
Good morning says I how do you do
But he proved a mighty rogue sirs
For at the corner of a lane
A pistol he pulls out sirs
And he rammed the muzzle arrah what a shame
Into my very mouth sirs

Your money hang your Irish eyes
Be merciful I cried sirs
He swore my brains he would blow out

319

Enlarged section of scrimshaw art on whale ivory
The men were sprawling in the sea
And swimming for their lives sir

LAST BUT NOT LEAST

If I should bawl or cry sirs
He levelled fair just for my sconce
Three steps I did retire sirs
His pistol flashed and his head I smashed
For a shillalah never misses fire

Euphrasia 1849

One suspects that this song was originally much longer than the version in the *Euphrasia* journal, but I have never seen it in print. I wish that more of the refrain had been recorded than just "Philliloo and etc." "Billy O'Rourke" sings very well to the melody for "The Girl I Left Behind Me," but I feel sure that it must have had a tune of its own.

NOW WE STEER OUR COURSE FOR HOME

The long sought time at length has come
And now we steer our course for home
Blow thou good wind and speed us on
The way that leads to our sweet home

Our ship seems conscious of the hour
That proves her strength and sailing power
She swiftly ploughs the parting tide
Her captain's and her seamen's pride

Our anxious friends methinks I hear
Repeat the day the month the year
Which tore us from their arms of love
O'er oceans's wide expanse to rove

With watchful eyes the spot they scan
Where stands the ready signal man
Who telegraphs each coming sail
Born forward by the favoring gale

Our hearts so long oppressed with care
Hasten on before to meet them there

While nightly in our dreams we trace
Each well known scene each loved one's face

We know that many a fervent prayer
They raise to ask our Father's care
Dash on old Uncas through the foam
To bear their sons and brothers home

Uncas 1843

The "ready signal man" in the fourth stanza was one who tele-graphed the arrival of in-bound ships by means of a relay of semaphore towers. There was one such tower on Tower Hill on Chappaquiddick, another on East Chop, and a third across the sound in Falmouth. From there the news was relayed on to Mattapoisett or Fairhaven or New Bed-ford, depending on the home port of the vessel.

INDEX OF SOURCES

THIS IS a list of the logbooks and journals which are the sources for the ballets used or referred to in this book. The following abbreviations are used:

Blosser; Gale Blosser, Milbrae, California.
Dartmouth; Old Dartmouth Historical Society, New Bedford, Mass.
Dukes; Dukes County Historical Society, Edgartown, Mass.
Essex; Essex Institute, Salem, Mass.
Hay; John Hay Library, Providence, R.I.
Huntington; Gale Huntington, Vineyard Haven, Mass.
Kendall; Kendall Whaling Museum, Sharon, Mass.
Mystic; Mystic Seaport Library, Mystic, Conn.
Nantucket; Nantucket Whaling Museum, Nantucket, Mass.
New Bedford; New Bedford Free Public Library, New Bedford, Mass.
Peabody; The Peabody Museum, Salem, Mass.
Pease; Mrs. Byron Pease, Nantucket, Mass.
Providence; Providence Public Library, Providence, R.I.
Smalley; Mrs. Amos Smalley, Gay Head, Mass.
Whiting; Everett W. Whiting, West Tisbury, Mass.

The entries in this Index of Sources are as follows: First, the name of the vessel with the rig prefixed. Second, the home port at the time of the voyage. Third, the year in which the recorded voyage began. And lastly, the repository in which the journal or logbook may be found.

Bark *Abraham Barker*, New Bedford, 1871. Hay.
Ship *Addison*, Tiverton, R. I., 1834, Providence.
Bank *Andrew Hicks*, New Bedford, 1879. Smalley.
Ship, *Ann*, Salem, 1776. Essex.
Brig *Argus*, Salem, 1813. Peabody.
Ship *Atkins Adams*, New Bedford, 1858. Dartmouth.
Ship *Bengal*, Salem, 1832. Providence (Journal of Ira Poland).

Ship *Bengal*, Salem, 1832. Essex. (Journal of William Silver).

Bark *Benjamin Cummins*, New Bedford, 1866. Dartmouth.

Ship *Benjamin Tucker*, New Bedford, 1849. Kendall.

Ship *Braganza*, New Bedford, 1845. Dartmouth.

Brig *Carthage*, Newburyport, 1842. Peabody.

Bark *Catalpa*, New Bedford, 1856. Providence.

Ship *Chile*, New Bedford, 1839. New Bedford.

Ship *Citizen*, Nantucket, 1844. Nantucket.

Ship *Clifford Wayne*, Nantucket, 1855. Nantucket.

Ship *Coral*, New Bedford, 1846. Dartmouth.

Ship *Cortes*, New Bedford, 1847. New Bedford. (A manuscript book of songs collected by William Histed.)

Ship, *Courier*, New Bedford, 1842. Dukes.

Ship *Dartmouth*, New Bedford, 1836. Providence.

Ship *Diana*, New York, 1819. Hay (Papers from a journal kept by Charles Murphy.)

Sloop *Dolphin*, Nantucket, 1790. Nantucket.

Ship *Edward*, New Bedford, 1849. Nantucket.

Ship *Edward Cary*, Nantucket, 1854. Nantucket.

Ship *Elbe*, Poughkeepsie, 1833. Blosser.

Ship *Eliza Adams*, New Bedford, 1879. Providence.

Ship *Elizabeth*, New Bedford, 1845. Providence.

Ship *Elizabeth*, New Bedford, 1847. Kendall.

Bark *Elizabeth Swift*, New Bedford, 1859. New Bedford.

Ship *Euphrasia*, New Bedford, 1849. Peabody.

Ship *Florida*, New Bedford, 1843. New Bedford.

Bark *Fortune*, Plymouth, 1840. Providence.

Ship *Frances Henrietta*, New Bedford, 1835. Huntington. (A manuscript book of songs and verse collected by Henry Manter.)

Ship *Galaxy*, Salem, 1827. Peabody.

Ship *George*, Salem, 1827. Peabody.

Ship *Gilmira*, New Bedford, 1835. Whiting.

Ship *Governor Carver*, Nantucket, 1854. Nantucket.

Ship *Herald*, Fairhaven, 1817. Providence.

Ship *Hercules*, New Bedford, 1828 Dartmouth.

Ship *Jasper*, Fairhaven, 1839. Providence.

Ship *Java*, New Bedford, 1839. Dartmouth.

Ship *Jirih Perry*, New Bedford, 1869. Dukes.

Sloop *Joseph Francis*, Boston, 1795. Peabody.

Bark *Josephine*, New Bedford, 1905. Dartmouth.

Ship *La Grange*, Salem, 1849. Peabody.

Ship *L. C. Richmond*, Salem, 1834. Essex.

Schooner, *Leopart*, Salem, 1767. Essex.

Ship *Lexington*, Nantucket, 1853. Nantucket.

Ship *Lotos*, Salem, 1833. Essex.

Ship *Marcus*, New Bedford, 1844. Whiting.

Ship *Maria*, Nantucket, 1832. Pease.

Ship *Maria*, Nantucket, 1848. Nantucket.

Ship *Mattapoisett*, Nantucket, 1852. Nantucket.

Bark *Midas*, New Bedford, 1861. Whiting.

Ship *Minerva*, New Bedford, 1845. New Bedford.

Ship *Minerva Smythe*, New Bedford, 1852. Dukes.

Ship *Nauticon*, New Bedford, 1848. Nantucket.

Ship *Nautilus*, New Bedford, 1838. Blosser.

Sloop *Nellie*, Edgartown, 1769. Dartmouth.

Bark *Ocean Rover*, New Bedford, 1859. Dartmouth.

Brigantine *Paulina*, Salem, 1808. Peabody.

Brig *Pavilion*, New Bedford, 1858. Whiting.

Brig *Polly*, Salem 1804. Essex.

Ship *Rebecca Simms*, New Bedford, 1851 Kendall.

Ship *Romulus*, Sag Harbor, 1851. Mystic

Ship *Sea Ranger*, New Bedford, 1879. Providence.

Ship *Sharon*, New Bedford, 1845. Dartmouth.

Bark *Smyrna*, New Bedford, 1853. Providence.

Bark *Stella*, New Bedford, 1860. Dartmouth.

Brig *Swan*, Salem, 1837. Peabody.

Ship *Thomas Perkins*, New York, 1844. Peabody.

Ship *Three Brothers*, Nantucket, 1846. Nantucket.

Ship *Trident*, New Bedford, 1846. Providence.

Brig *Two Brothers*, Wethersfield, 1768. Mystic.

Ship *Uncas*, New Bedford, 1843. Providence.

Ship *Vaughn*, Salem, 1767. Essex.

Ship *Walter Scott*, Nantucket, 1840. Nantucket.

Ship *Young Phoenix*, New Bedford, 1844. New Bedford.

BIBLIOGRAPHY

The American Musical Miscellany, Northampton, Mass., 1798.

Bantock, Granville: *One Hundred Songs of England,* New York, Oliver Ditson Co., 1914.

Baring-Gould, Rev. Sabine, and H. Fleetwood Sheppard: *A Garland Of Country Song,* London, Methuen & Co., 1895.

Barrett, William Alexander: *English Folk Songs,* London, H. W. Gray Co., 1891.

Barry, Phillips: *The Maine Woods Songster,* Cambridge, Mass., The Powell Printing Co., 1939.

Barry, Phillips, Fannie H. Eckstrom, and Mary W. Smyth: *British Ballads From Maine,* New Haven, The Yale University Press, 1929.

Blake, Charles D.: *Harmonized Melodies, No. 34,* Boston, F. Trifet, 1893.

Bone, David W.; *Capstan Bars,* Edinburgh, The Porpoise Press, 1931.

The Book Of Navy Songs, Garden City, N. Y., Doubleday Doran & Co., 1930.

Broadwood, Lucy E.: *English Traditional Songs and Carols,* London, Boosey & Hawks, 1909.

Broadwood, Lucy E., and J. A. Fuller Maitland: *English County Songs,* J. B. Cramer & Co., Ltd., n.d.

Bulletin Of The Folksong Society Of The Northeast, Philadelphia, Reprint, The American Folklore Society, Inc., 1960.

Bunting: *Music Of Ireland,* Dublin, Hodges & Son, 1840.

Calliope, London, 1788.

Cazden, Norman: *The Abelard Folk Song Book,* New York, Abelard Schuman, 1958.

Chappell, William: *A Collection Of National English Airs,* 2 vols., London, Chappell, 1840.

Clements, Rex: *Manavillins: A Muster of English Sea-Songs,* London, Heath, Cranton, Ltd., 1928.

The Clown's Songster, New York, Robt. M. DeWitt, 1871.

Colcord, Joanna C.: *Songs Of American Sailormen,* New York, W. W. Norton Co., Inc., 1938.

Creighton, Helen: *Songs and Ballads From Nova Scotia,* Toronto, J. M. Dent & Sons, Ltd., 1933. Dover reprint, 1966.

Creighton, Helen, and Doreen H. Senior: *Traditional Songs From Nova Scotia,* Toronto, The Ryerson Press, 1950.

Creighton, Helen: *Maritime Folk Songs,* Toronto, The Ryerson Press, 1961.

Dolph, Edward Arthur: *Sound Off,* New York, Farrar & Rinehart Inc., 1942.

Downs, Olin, and Elie Siegmeister: *A Treasury Of American Song,* New York, Howell Soskin & Co., 1940.

Doerflinger, William: *Shanteymen And Shanty Boys,* New York, MacMillan Co., 1951.

Duncan, Edmondstoune: *The Minstrelsy Of England,* London, Augener, Ltd., 1905.

Flanders, Helen Hartness and Marguerite Olney: *Ballads Migrant In New England,* New York, Farrar Straus and Young, 1953.

BIBLIOGRAPHY

Flanders, Helen Hartness: *Ancient Ballads Traditionally Sung In New England,* Philadelphia, The University of Pennsylvania Press, 1961.

Ford, Robert: *Vagabond Songs And Ballads Of Scotland,* 2 vols., London, Alexander Gardner, 1899, 1901.

Fuson, Harvey H.: *Ballads Of The Kentucky Highlands,* London, The Mitre Press, 1931.

Galvin, Patrick: *Irish Songs Of Resistance,* New York, The Folklore Press, n.d.

Gems of Scottish Song, Boston, Oliver Ditson & Co., 1866.

The Golden Wreath, Boston, Oliver Ditson & Co., 1857.

Greenleaf, Elizabeth Bristol, and Grace Yarrow Mansfield: *Ballads And Sea Songs From Newfoundland,* Cambridge, Mass., Harvard University Press, 1933:

Harlow, Frederick Pease: *Chanteying Aboard American Ships,* Barre, Mass., The Barre Publishing Co., 1962. Dover reprint, 1970.

Hatton, J. L.: *Songs of England,* London, Boosey & Co., 1888.

Henry, Sam: *Songs Of The People,* a manuscript and newspaper-clipping collection of songs in the National Library of Ireland, Dublin, collected between 1923 and 1939.

Howe's 100 Old Favorite Songs, Boston, Elias Howe Co., n.d.

Hudson, Henry: A manuscript collection of Irish melodies, 5 vols., in the Boston Public Library, Boston, Mass., collected c. 1840.

Hugil, Stan: *Shanties From The Seven Seas,* New York, E. P. Dutton & Co., 1961.

Johnson, H. K.: *Our Familiar Songs,* New York, H. Holt, 1881.

The Journal Of American Folklore, 1888-

Journal Of the Folk-Song Society, 1899-1931.

Joyce, Patrick Weston: *Ancient Irish Music,* Dublin, M. H. Gill, 1890.

Joyce, Patrick Weston: *Old Irish Music,* London, Longmans Green & Co., 1909.

Kidson, Frank: *Traditional Tunes,* Oxford, Taphouse & Son, 1891.

Kidson, Frank, and Alfred Moffat: *A Garland Of English Folk-Songs,* London, Ascherberg Hopwood & Crew, Ltd., n.d.

Kidson, Frank and Ethel Kidson, ed.; *English Peasant Songs,* London, 1929.

Kidson, Frank: *Songs of Great Britain,* New York, Boosey & Co., 1913.

Kincaid, Bradley: *My Favorite Mountain Ballads And Old Time Songs,* Chicago, 1928.

Kincaid, Bradley: *Favorite Old Time Songs . . . No. 2,* Chicago, 1929.

Kincaid, Bradley: *Favorite Old Time Songs . . . No. 3,* Chicago, 1930.

Kitchiner, William: *Sea Songs Of England,* London, Hurst, Robinson & Co., 1823.

Laurel Song Book, Boston, Birchard & Co., 1900.

Linscott, Eloise Hubbard: *Folk Songs Of Old New England,* New York, The MacMillan Co., 1939.

Litten, William: A manuscript collection of fiddle tunes, in the library of the Dukes County Historical Society, Inc., Edgartown, Mass., collected in 1800-1802.

Lochlainn, Colm O.: *Irish Street Ballads,* New York, Corinth Books, Inc., 1960.

Lomax, John A., and Alan Lomax: *American Ballads And Folk Songs,* New York, The Macmillan Co., 1934.

Lomax, John A., and Alan Lomax: *Best Loved American Folk Songs,* New York, Grosset & Dunlap, 1947.

Mackay, Charles: *The Songs Of Scotland,* Boston, Oliver Ditson Co., 1888.

Mackenzie, W. Roy: *Ballads And Sea Songs From Nova Scotia,* Cambridge, Mass., 1928.

The Mammoth Songster, Boston, Birchard & Co., 1866.

McCaskey, J. P.: *The Franklin Square Song Collections,* 8 vols., New York, Harper & Brothers, 1881-1892.

Moffat, Alfred: *The Minstrelsy Of England,* London, Bayley & Ferguson, n.d.

The Mohawk Minstrel's Magazine, London, Francis Day & Hunter, 1878.

O'Neil, Francis: *Music Of Ireland*, Chicago, 1903.

Petrie, George: *The Complete Collection Of Irish Music*, London, Boosey & Co., 1902.

Sandburg, Carl: *The American Songbag*, New York, Harcourt Brace & Co., 1927.

Sharp, Cecil J.: *English Folk Songs From The Southern Appalachians*, reprint, 2 vols. in 1, London, Oxford University Press, 1960.

Sharp, Cecil J.: *One Hundred English Folk Songs*, New York, 1916.

Sharp, Cecil J.: *English Folk Songs*, reprint, 2 vols. in 1, London, Novello & Co., 1959.

The Silver Chord, Boston, Oliver Ditson & Co., 1862.

Smith, Laura: *The Music Of The Waters*, London, Kegan Paul & Co., 1888.

Songs That Never Die, 1894.

The Songster's Museum, Northampton, Mass., 1803.

Spaeth, Sigmund: *Read 'Em And Weep*, Garden City, N. Y., Doubleday Page & Co., 1926.

Spaeth, Sigmund: *Weep Some More My Lady*, Garden City, N.Y., Doubleday Page & Co., 1927.

Stone, Christopher: *Sea Songs And Ballads*, Oxford, The Clarendon Press 1906.

Vaughan Williams, Ralph: *Folk Songs From The Eastern Counties*, London, 1908.

Vaughan, Williams, Ralph, and A. L. Lloyd: *The Penguin Book Of English Folk Songs*, London, Penguin Books, 1959.

The Violin Players Pastime, New York, Carl Fischer, n.d.

Whall, W. B.: *Sea Songs And Shanties*, Glasgow, Brown, Son & Ferguson, Ltd., 1948.

Wells, Evelyn Kendrick: *The Ballad Tree*, New York, The Ronald Press, 1950.

Winner, Sep: *World Of Song*, New York, N. D. Thompson, 1883.

White's Unique Collection Of Jigs, Reels, Etc., New York, The White-Smith Music Publishing Co., 1896.

Williams, Alfred: *Folk-Songs Of the Upper Thames*, London, Duckworth & Co., 1923.

Williams, Alfred M.: *Studies In Folk-Song And Popular Poetry*, Boston, Houghton Mifflin & Co., 1894.

Wilson, James: *The Musical Cyclopedia*, London, Bell & Co., 1834.

Wyman, Loraine: *Twenty Kentucky Mountain Songs*, Boston, Oliver Ditson Co., 1920.

INDEX OF SONG TITLES

Adieu My Native Land, 238

Adieu to Erin, 255

American Stranger, The. *See* When First into This Country

Ancient Riddle, An, 282

Angels Whisper, 239

Aran's Lonely Home, 198

As I Grow Old, 300

Banks of Banna, The, 236

Banks of Champlain, The, 161

Banks of Glenco, The, 113

Banks of the Schuylkill, The, 160

Bark Gay Head, The, 34

Bark Ocean Rover, The, 37

Barney McCoy. *See* Norah Darling

Beacon Light, The, 260

Beggarman, The, 116

Behind the Green Bush, 223

Betsy Is a Beauty Fair. *See* Fair Betsy

Bible Story, The, 264

Billy O'Rourke, 318

Blessed Land of Love and Liberty, 170

Blow High Blow Low, 256

Blow Ye Winds, 42

Bold Privateer, The. *See* The Captain Calls All Hands

Bonaparte, 209

Bonaparte on St. Helena, 205

Bonnet of Blue, The, 275

Bonny Bunch of Roses-O, The, 207

Bride's Farewell, The, 241

Bright Phoebe, 119

British Man-of-War, The, 108

California Song, The, 174

Can of Grog, The, 73

Captain, The, 176

Captain Calls All Hands, The, 99

Captain James, 54

Captain's Apprentice, The. *See* Captain James

Caroline and Her Young Sailor Bold. *See* The Nobleman's Daughter

Caroline of Edinburgh Town. *See* Lovely Caroline

Charming Fellow, A, 296

Coast of Peru, The, 2

Come Let Us Be Jolly, 277

Confession, The, 179

County of Tyrone, The, 218

Covent Garden, 90

Croppy Boy, The, 188

Cruise of the Dove, The, 13

Cupid's Garden, 92

Cupid's Garden. *See* Covent Garden

Dark-Eyed Sailor, The, 120

Dauntless Sailor, The, 85

Demon of the Sea, The, 78

Desolation, 38

Diego's Bold Shore, 30

Down Wapping, 263

Dying Soldier, The, 243

Elegy on the Death of a Mad Dog, 295

Erin's Lovely Home. *See* Aran's Lonely Home

Factor's Song, The. *See* The Turkey Factor in Foreign Parts

Fair Betsy, 201

Fanny Blair, 229

Fare You Well, 274

Farewell My Dear Nancy, 266

Farmer's Boy, A, 216

First Time I Saw My Love, The, 225

Fitting Out, A, 7

Flora the Lily of the West. *See* The Lily of the West

Flowing Bowl, The. *See* Sling the Flowing Bowl

Genette and Genoe, 245

Gossport Beach. *See* The Undutiful Daughter

Gossport Tragedy, The. *See* The Ship Carpenter

Green Linnet, The, 211

Greenland Whale, The, 11

Hearts of Gold, 68

Heathen Dear, The, 151

Hunter's Lane, 273

I Can Not Call Her Mother, 298

I Had a Handsome Fortune, 203

I Was Once a Sailor, 66

In Days When We Went Gipsying, 220

Indian Hunter, The, 180

Jacket So Blue, The. *See* The Bonnet of Blue

Jamie's on the Stormy Sea, 252

John Brown, 158

John Bull's Epistle, 172

John Riley, 105

Keyhole in the Door, The, 315

Lass of Mowee, The, 148

Life on the Ocean Wave, A, 87

Lily of the West, The, 133

Little Mary the Sailor's Bride. *See* The Beggarman

Little Mohee, The. *See* The Lass of Mowee

Loose Every Sail to the Breeze, 52

Lord Our God, The, 289

Love Song in the Year 1769, A, 293

Lovely Caroline, 137

Maid of Erin, The, 237

Maid on the Shore, The, 136

Mantle So Green, The, 122

Mary's Dream, 246

McDonald's Return to Glenco. *See* The Banks of Glenco

Moll Brooks, 274

Moon Is Brightly Beaming Love, The, 152

Most Beautiful, 81

Mother's Admonition, The. *See* The Tarry Trousers

My Flora and I. *See* The Shepherd's Lament

Nelson, 275

Neptune, 83

New Liberty Song, A, 163

New Sea Song, A, 262

New Song, A, 187

Nobleman's Daughter, The, 103

Norah Darling, 302

Now We Steer Our Course for Home, 321

O Logie O Buchan, 197

Ocean, The, 248

Ocean Queen, The, 82

Old Horse, 279

Old Hulk, The, 32

One Night Sad and Languid, 215

Our Old Friend Coffin, 22

Our Ship She Is Lying in Harbour, 124

Phoebe. *See* Bright Phoebe

Pilot, The, 288

Pirate of the Isles, The, 74

Plowboy's Courtship, The. *See* Queen of the May

Poll and Sal, 312

Post Below, The, 292

Prayer, 286

Prayer at the Start of a Voyage, A, 7

Pretty Sally, 111

Pride of Kildare, The, 131

Queen of the May, 190

Recruiting Sargeant, The, 291

Reily's Jailed, 224

Rinordine, 222

Rolling Down to Old Mohee, 27

Rose of Allendale, The, 257

Rover of the Sea, The, 80

Row On, 290

Sailor Boy's Song, The, 271

Sailor from Dover, The. *See* Pretty Sally

A CATALOGUE OF SELECTED DOVER BOOKS
IN ALL FIELDS OF INTEREST

WHAT IS SCIENCE?, *N. Campbell*
The role of experiment and measurement, the function of mathematics, the nature of scientific laws, the difference between laws and theories, the limitations of science, and many similarly provocative topics are treated clearly and without technicalities by an eminent scientist. "Still an excellent introduction to scientific philosophy," H. Margenau in *Physics Today*. "A first-rate primer . . . deserves a wide audience," *Scientific American*. 192pp. 5⅜ x 8.
60043-2 Paperbound $1.25

THE NATURE OF LIGHT AND COLOUR IN THE OPEN AIR, *M. Minnaert*
Why are shadows sometimes blue, sometimes green, or other colors depending on the light and surroundings? What causes mirages? Why do multiple suns and moons appear in the sky? Professor Minnaert explains these unusual phenomena and hundreds of others in simple, easy-to-understand terms based on optical laws and the properties of light and color. No mathematics is required but artists, scientists, students, and everyone fascinated by these "tricks" of nature will find thousands of useful and amazing pieces of information. Hundreds of observational experiments are suggested which require no special equipment. 200 illustrations; 42 photos. xvi + 362pp. 5⅜ x 8.
20196-1 Paperbound $2.00

THE STRANGE STORY OF THE QUANTUM, AN ACCOUNT FOR THE GENERAL READER OF THE GROWTH OF IDEAS UNDERLYING OUR PRESENT ATOMIC KNOWLEDGE, *B. Hoffmann*
Presents lucidly and expertly, with barest amount of mathematics, the problems and theories which led to modern quantum physics. Dr. Hoffmann begins with the closing years of the 19th century, when certain trifling discrepancies were noticed, and with illuminating analogies and examples takes you through the brilliant concepts of Planck, Einstein, Pauli, Broglie, Bohr, Schroedinger, Heisenberg, Dirac, Sommerfeld, Feynman, etc. This edition includes a new, long postscript carrying the story through 1958. "Of the books attempting an account of the history and contents of our modern atomic physics which have come to my attention, this is the best," H. Margenau, Yale University, in *American Journal of Physics*. 32 tables and line illustrations. Index. 275pp. 5⅜ x 8.
20518-5 Paperbound $2.00

GREAT IDEAS OF MODERN MATHEMATICS: THEIR NATURE AND USE, *Jagjit Singh*
Reader with only high school math will understand main mathematical ideas of modern physics, astronomy, genetics, psychology, evolution, etc. better than many who use them as tools, but comprehend little of their basic structure. Author uses his wide knowledge of non-mathematical fields in brilliant exposition of differential equations, matrices, group theory, logic, statistics, problems of mathematical foundations, imaginary numbers, vectors, etc. Original publication. 2 appendixes. 2 indexes. 65 ills. 322pp. 5⅜ x 8.
20587-8 Paperbound $2.25

Sailor's Adieu, The. *See* The Topsail Shivers in the Wind

Sailor's Come All Ye, The. *See* Hearts of Gold

Sailor's Farewell, The, 304

Sailor's Trade Is a Roving Life, A, 272

Sandshark, The, 317

Sarah Mariah Cornell, 156

Saturday Night at Sea, 65

Sea, The, 63

Sea Captain, The. *See* The Maid on the Shore

Sea Ran High, The, 81

Sequel to Will Watch, The, 62

Sheffield Prentice Boy, The, 192

Shepherd's Daughter, The, 185

Shepherd's Lament, The, 227

Shepherd's Resolution, The. *See* Behind the Green Bush

Ship Carpenter, The, 129

Ship Euphrasia, The, 47

Silvery Moon, 233

Silvery Tide, The, 125

Sling the Flowing Bowl, 51

Song of Solomon's Temple, 309

Song on Courtship, 194

Song on the Nantucket Ladies, A, 165

Sons of Liberty, The, 146

Sons of Worth, 171

Sovereign of the Sea, The, 85

Springfield Mountain, 167

Storm, The. *See* The Tempest

Storm Was Loud, The, 83

Susan the Pride of Kildare. *See* The Pride of Kildare

Tarry Trousers, The, 96

Tempest, The, 70

Terrible Polly, 306

There She Blows. *See* The Wounded Whale

Thou Hast Learned to Love Another, 249

Times, The, 144

Topsail Shivers in the Wind, The, 59

Turkey Factor in Foreign Parts, The, 268

Turkish Lady, The, 141

Undutiful Daughter, The, 127

Village Born Beauty, The, 299

Virtuous America, 180

Wait for the Wagon, 285

We Met 'Twas in a Crowd, 251

Wedding Rite, The. *See* I Can Not Call Her Mother

Wet Sheet and a Flowing Sea, A, 49

Whalefish Song, The, 9

Whaleman's Lament, The, 15

Whalers' Song, The, 17

Whaling Scene, A, 21

When First into this Country, 195

When I Remember, 277

Wide World of Waters, The, 308

William Taylor, 94

Willie and Mary. *See* The Beggarman

Willie Gray, 182

Willie's on the Dark Blue Sea, 234

Willy Rily. *See* Reily's Jailed

Wings of a Goney, The, 40

Women Love Kissing as Well as the Men, 228

Wounded Whale, The, 23

Wreath, The, 297

Yankee Doodle. *See* The Times

Ye Parliaments of England, 278

Yonder Stands a Handsome Lady. *See* Song on Courtship

Young Sailor Bold, The. *See* The Nobleman's Daughter

Young Shepherd, The. *See* The Shepherd's Lament

Young Virgin, A, 100

A CATALOGUE OF SELECTED DOVER BOOKS
IN ALL FIELDS OF INTEREST

THE MUSIC OF THE SPHERES: THE MATERIAL UNIVERSE — FROM ATOM TO QUASAR, SIMPLY EXPLAINED, *Guy Murchie*
Vast compendium of fact, modern concept and theory, observed and calculated data, historical background guides intelligent layman through the material universe. Brilliant exposition of earth's construction, explanations for moon's craters, atmospheric components of Venus and Mars (with data from recent fly-by's), sun spots, sequences of star birth and death, neighboring galaxies, contributions of Galileo, Tycho Brahe, Kepler, etc.; and (Vol. 2) construction of the atom (describing newly discovered sigma and xi subatomic particles), theories of sound, color and light, space and time, including relativity theory, quantum theory, wave theory, probability theory, work of Newton, Maxwell, Faraday, Einstein, de Broglie, etc. "Best presentation yet offered to the intelligent general reader," *Saturday Review*. Revised (1967). Index. 319 illustrations by the author. Total of xx + 644pp. 5⅜ x 8½.
21809-0, 21810-4 Two volume set, paperbound $5.00

FOUR LECTURES ON RELATIVITY AND SPACE, *Charles Proteus Steinmetz*
Lecture series, given by great mathematician and electrical engineer, generally considered one of the best popular-level expositions of special and general relativity theories and related questions. Steinmetz translates complex mathematical reasoning into language accessible to laymen through analogy, example and comparison. Among topics covered are relativity of motion, location, time; of mass; acceleration; 4-dimensional time-space; geometry of the gravitational field; curvature and bending of space; non-Euclidean geometry. Index. 40 illustrations. x + 142pp. 5⅜ x 8½.
61771-8 Paperbound $1.35

HOW TO KNOW THE WILD FLOWERS, *Mrs. William Starr Dana*
Classic nature book that has introduced thousands to wonders of American wild flowers. Color-season principle of organization is easy to use, even by those with no botanical training, and the genial, refreshing discussions of history, folklore, uses of over 1,000 native and escape flowers, foliage plants are informative as well as fun to read. Over 170 full-page plates, collected from several editions, may be colored in to make permanent records of finds. Revised to conform with 1950 edition of Gray's Manual of Botany. xlii + 438pp. 5⅜ x 8½.
20332-8 Paperbound $2.50

MANUAL OF THE TREES OF NORTH AMERICA, *Charles Sprague Sargent*
Still unsurpassed as most comprehensive, reliable study of North American tree characteristics, precise locations and distribution. By dean of American dendrologists. Every tree native to U.S., Canada, Alaska; 185 genera, 717 species, described in detail—leaves, flowers, fruit, winterbuds, bark, wood, growth habits, etc. plus discussion of varieties and local variants, immaturity variations. Over 100 keys, including unusual 11-page analytical key to genera, aid in identification. 783 clear illustrations of flowers, fruit, leaves. An unmatched permanent reference work for all nature lovers. Second enlarged (1926) edition. Synopsis of families. Analytical key to genera. Glossary of technical terms. Index. 783 illustrations, 1 map. Total of 982pp. 5⅜ x 8.
20277-1, 20278-X Two volume set, paperbound $6.00

IT'S FUN TO MAKE THINGS FROM SCRAP MATERIALS,
Evelyn Glantz Hershoff
What use are empty spools, tin cans, bottle tops? What can be made from
rubber bands, clothes pins, paper clips, and buttons? This book provides
simply worded instructions and large diagrams showing you how to make
cookie cutters, toy trucks, paper turkeys, Halloween masks, telephone sets,
aprons, linoleum block- and spatter prints — in all 399 projects! Many are easy
enough for young children to figure out for themselves; some challenging
enough to entertain adults; all are remarkably ingenious ways to make things
from materials that cost pennies or less! Formerly "Scrap Fun for Everyone."
Index. 214 illustrations. 373pp. 5⅜ x 8½.   21251-3 Paperbound $1.75

SYMBOLIC LOGIC and THE GAME OF LOGIC, *Lewis Carroll*
"Symbolic Logic" is not concerned with modern symbolic logic, but is instead
a collection of over 380 problems posed with charm and imagination, using
the syllogism and a fascinating diagrammatic method of drawing conclusions.
In "The Game of Logic" Carroll's whimsical imagination devises a logical game
played with 2 diagrams and counters (included) to manipulate hundreds of
tricky syllogisms. The final section, "Hit or Miss" is a lagniappe of 101 additional puzzles in the delightful Carroll manner. Until this reprint edition,
both of these books were rarities costing up to $15 each. Symbolic Logic:
Index. xxxi + 199pp. The Game of Logic: 96pp. 2 vols. bound as one. 5⅜ x 8.
                                                          20492-8 Paperbound $2.50

MATHEMATICAL PUZZLES OF SAM LOYD, PART I
selected and edited by M. Gardner
Choice puzzles by the greatest American puzzle creator and innovator. Selected
from his famous collection, "Cyclopedia of Puzzles," they retain the unique
style and historical flavor of the originals. There are posers based on arithmetic,
algebra, probability, game theory, route tracing, topology, counter and sliding
block, operations research, geometrical dissection. Includes the famous "14-15"
puzzle which was a national craze, and his "Horse of a Different Color" which
sold millions of copies. 117 of his most ingenious puzzles in all. 120 line
drawings and diagrams. Solutions. Selected references. xx + 167pp. 5⅜ x 8.
                                                          20498-7 Paperbound $1.35

STRING FIGURES AND HOW TO MAKE THEM, *Caroline Furness Jayne*
107 string figures plus variations selected from the best primitive and modern
examples developed by Navajo, Apache, pygmies of Africa, Eskimo, in Europe,
Australia, China, etc. The most readily understandable, easy-to-follow book in
English on perennially popular recreation. Crystal-clear exposition; step-by-
step diagrams. Everyone from kindergarten children to adults looking for
unusual diversion will be endlessly amused. Index. Bibliography. Introduction
by A. C. Haddon. 17 full-page plates, 960 illustrations. xxiii + 401pp. 5⅜ x 8½.
                                                          20152-X Paperbound $2.25

PAPER FOLDING FOR BEGINNERS, *W. D. Murray and F. J. Rigney*
A delightful introduction to the varied and entertaining Japanese art of
origami (paper folding), with a full, crystal-clear text that anticipates every
difficulty; over 275 clearly labeled diagrams of all important stages in creation.
You get results at each stage, since complex figures are logically developed
from simpler ones. 43 different pieces are explained: sailboats, frogs, roosters,
etc. 6 photographic plates. 279 diagrams. 95pp. 5⅜ x 8⅜.
                                                          20713-7 Paperbound $1.00

PRINCIPLES OF ART HISTORY,
H. Wölfflin
Analyzing such terms as "baroque," "classic," "neoclassic," "primitive," "picturesque," and 164 different works by artists like Botticelli, van Cleve, Dürer, Hobbema, Holbein, Hals, Rembrandt, Titian, Brueghel, Vermeer, and many others, the author establishes the classifications of art history and style on a firm, concrete basis. This classic of art criticism shows what really occurred between the 14th-century primitives and the sophistication of the 18th century in terms of basic attitudes and philosophies. "A remarkable lesson in the art of seeing," *Sat. Rev. of Literature*. Translated from the 7th German edition. 150 illustrations. 254pp. 6⅛ x 9¼. 20276-3 Paperbound $2.25

PRIMITIVE ART,
Franz Boas
This authoritative and exhaustive work by a great American anthropologist covers the entire gamut of primitive art. Pottery, leatherwork, metal work, stone work, wood, basketry, are treated in detail. Theories of primitive art, historical depth in art history, technical virtuosity, unconscious levels of patterning, symbolism, styles, literature, music, dance, etc. A must book for the interested layman, the anthropologist, artist, handicrafter (hundreds of unusual motifs), and the historian. Over 900 illustrations (50 ceramic vessels, 12 totem poles, etc.). 376pp. 5⅜ x 8. 20025-6 Paperbound $2.50

THE GENTLEMAN AND CABINET MAKER'S DIRECTOR,
Thomas Chippendale
A reprint of the 1762 catalogue of furniture designs that went on to influence generations of English and Colonial and Early Republic American furniture makers. The 200 plates, most of them full-page sized, show Chippendale's designs for French (Louis XV), Gothic, and Chinese-manner chairs, sofas, canopy and dome beds, cornices, chamber organs, cabinets, shaving tables, commodes, picture frames, frets, candle stands, chimney pieces, decorations, etc. The drawings are all elegant and highly detailed; many include construction diagrams and elevations. A supplement of 24 photographs shows surviving pieces of original and Chippendale-style pieces of furniture. Brief biography of Chippendale by N. I. Bienenstock, editor of *Furniture World*. Reproduced from the 1762 edition. 200 plates, plus 19 photographic plates. vi + 249pp. 9⅛ x 12¼. 21601-2 Paperbound $3.50

AMERICAN ANTIQUE FURNITURE: A BOOK FOR AMATEURS,
Edgar G. Miller, Jr.
Standard introduction and practical guide to identification of valuable American antique furniture. 2115 illustrations, mostly photographs taken by the author in 148 private homes, are arranged in chronological order in extensive chapters on chairs, sofas, chests, desks, bedsteads, mirrors, tables, clocks, and other articles. Focus is on furniture accessible to the collector, including simpler pieces and a larger than usual coverage of Empire style. Introductory chapters identify structural elements, characteristics of various styles, how to avoid fakes, etc. "We are frequently asked to name some book on American furniture that will meet the requirements of the novice collector, the beginning dealer, and . . . the general public. . . . We believe Mr. Miller's two volumes more completely satisfy this specification than any other work," *Antiques*. Appendix. Index. Total of vi + 1106pp. 7⅞ x 10¾. 21599-7, 21600-4 Two volume set, paperbound $7.50

THE BAD CHILD'S BOOK OF BEASTS, MORE BEASTS FOR WORSE CHILDREN, and A MORAL ALPHABET, *H. Belloc*
Hardly and anthology of humorous verse has appeared in the last 50 years without at least a couple of these famous nonsense verses. But one must see the entire volumes — with all the delightful original illustrations by Sir Basil Blackwood — to appreciate fully Belloc's charming and witty verses that play so subacidly on the platitudes of life and morals that beset his day — and ours. A great humor classic. Three books in one. Total of 157pp. 5⅜ x 8.
20749-8 Paperbound $1.00

THE DEVIL'S DICTIONARY, *Ambrose Bierce*
Sardonic and irreverent barbs puncturing the pomposities and absurdities of American politics, business, religion, literature, and arts, by the country's greatest satirist in the classic tradition. Epigrammatic as Shaw, piercing as Swift, American as Mark Twain, Will Rogers, and Fred Allen, Bierce will always remain the favorite of a small coterie of enthusiasts, and of writers and speakers whom he supplies with "some of the most gorgeous witticisms of the English language" (H. L. Mencken). Over 1000 entries in alphabetical order. 144pp. 5⅜ x 8.
20487-1 Paperbound $1.00

THE COMPLETE NONSENSE OF EDWARD LEAR.
This is the only complete edition of this master of gentle madness available at a popular price. *A Book of Nonsense, Nonsense Songs, More Nonsense Songs and Stories* in their entirety with all the old favorites that have delighted children and adults for years. The Dong With A Luminous Nose, The Jumblies, The Owl and the Pussycat, and hundreds of other bits of wonderful nonsense. 214 limericks, 3 sets of Nonsense Botany, 5 Nonsense Alphabets, 546 drawings by Lear himself, and much more. 320pp. 5⅜ x 8. 20167-8 Paperbound $1.75

THE WIT AND HUMOR OF OSCAR WILDE, *ed. by Alvin Redman*
Wilde at his most brilliant, in 1000 epigrams exposing weaknesses and hypocrisies of "civilized" society. Divided into 49 categories—sin, wealth, women, America, etc.—to aid writers, speakers. Includes excerpts from his trials, books, plays, criticism. Formerly "The Epigrams of Oscar Wilde." Introduction by Vyvyan Holland, Wilde's only living son. Introductory essay by editor. 260pp. 5⅜ x 8.
20602-5 Paperbound $1.50

A CHILD'S PRIMER OF NATURAL HISTORY, *Oliver Herford*
Scarcely an anthology of whimsy and humor has appeared in the last 50 years without a contribution from Oliver Herford. Yet the works from which these examples are drawn have been almost impossible to obtain! Here at last are Herford's improbable definitions of a menagerie of familiar and weird animals, each verse illustrated by the author's own drawings. 24 drawings in 2 colors; 24 additional drawings. vii + 95pp. 6½ x 6.   21647-0 Paperbound $1.00

THE BROWNIES: THEIR BOOK, *Palmer Cox*
The book that made the Brownies a household word. Generations of readers have enjoyed the antics, predicaments and adventures of these jovial sprites, who emerge from the forest at night to play or to come to the aid of a deserving human. Delightful illustrations by the author decorate nearly every page. 24 short verse tales with 266 illustrations. 155pp. 6⅝ x 9¼.
21265-3 Paperbound $1.50

THE PRINCIPLES OF PSYCHOLOGY,
William James

The full long-course, unabridged, of one of the great classics of Western literature and science. Wonderfully lucid descriptions of human mental activity, the stream of thought, consciousness, time perception, memory, imagination, emotions, reason, abnormal phenomena, and similar topics. Original contributions are integrated with the work of such men as Berkeley, Binet, Mills, Darwin, Hume, Kant, Royce, Schopenhauer, Spinoza, Locke, Descartes, Galton, Wundt, Lotze, Herbart, Fechner, and scores of others. All contrasting interpretations of mental phenomena are examined in detail—introspective analysis, philosophical interpretation, and experimental research. "A classic," *Journal of Consulting Psychology*. "The main lines are as valid as ever," *Psychoanalytical Quarterly*. "Standard reading . . . a classic of interpretation," *Psychiatric Quarterly*. 94 illustrations. 1408pp. 5⅜ x 8.

20381-6, 20382-4 Two volume set, paperbound $6.00

VISUAL ILLUSIONS: THEIR CAUSES, CHARACTERISTICS AND APPLICATIONS,
M. Luckiesh

"Seeing is deceiving," asserts the author of this introduction to virtually every type of optical illusion known. The text both describes and explains the principles involved in color illusions, figure-ground, distance illusions, etc. 100 photographs, drawings and diagrams prove how easy it is to fool the sense: circles that aren't round, parallel lines that seem to bend, stationary figures that seem to move as you stare at them — illustration after illustration strains our credulity at what we see. Fascinating book from many points of view, from applications for artists, in camouflage, etc. to the psychology of vision. New introduction by William Ittleson, Dept. of Psychology, Queens College. Index. Bibliography. xxi + 252pp. 5⅜ x 8½.

21530-X Paperbound $1.50

FADS AND FALLACIES IN THE NAME OF SCIENCE,
Martin Gardner

This is the standard account of various cults, quack systems, and delusions which have masqueraded as science: hollow earth fanatics. Reich and orgone sex energy, dianetics, Atlantis, multiple moons, Forteanism, flying saucers, medical fallacies like iridiagnosis, zone therapy, etc. A new chapter has been added on Bridey Murphy, psionics, and other recent manifestations in this field. This is a fair, reasoned appraisal of eccentric theory which provides excellent inoculation against cleverly masked nonsense. "Should be read by everyone, scientist and non-scientist alike," R. T. Birge, Prof. Emeritus of Physics, Univ. of California; Former President, American Physical Society. Index. x + 365pp. 5⅜ x 8.

20394-8 Paperbound $2.00

ILLUSIONS AND DELUSIONS OF THE SUPERNATURAL AND THE OCCULT,
D. H. Rawcliffe

Holds up to rational examination hundreds of persistent delusions including crystal gazing, automatic writing, table turning, mediumistic trances, mental healing, stigmata, lycanthropy, live burial, the Indian Rope Trick, spiritualism, dowsing, telepathy, clairvoyance, ghosts, ESP, etc. The author explains and exposes the mental and physical deceptions involved, making this not only an exposé of supernatural phenomena, but a valuable exposition of characteristic types of abnormal psychology. Originally titled "The Psychology of the Occult." 14 illustrations. Index. 551pp. 5⅜ x 8. 20503-7 Paperbound $3.50

FAIRY TALE COLLECTIONS, *edited by Andrew Lang*
Andrew Lang's fairy tale collections make up the richest shelf-full of traditional children's stories anywhere available. Lang supervised the translation of stories from all over the world—familiar European tales collected by Grimm, animal stories from Negro Africa, myths of primitive Australia, stories from Russia, Hungary, Iceland, Japan, and many other countries. Lang's selection of translations are unusually high; many authorities consider that the most familiar tales find their best versions in these volumes. All collections are richly decorated and illustrated by H. J. Ford and other artists.

THE BLUE FAIRY BOOK. 37 stories. 138 illustrations. ix + 390pp. 5⅜ x 8½.
21437-0 Paperbound $1.95

THE GREEN FAIRY BOOK. 42 stories. 100 illustrations. xiii + 366pp. 5⅜ x 8½.
21439-7 Paperbound $1.75

THE BROWN FAIRY BOOK. 32 stories. 50 illustrations, 8 in color. xii + 350pp. 5⅜ x 8½.
21438-9 Paperbound $1.95

THE BEST TALES OF HOFFMANN, *edited by E. F. Bleiler*
10 stories by E. T. A. Hoffmann, one of the greatest of all writers of fantasy. The tales include "The Golden Flower Pot," "Automata," "A New Year's Eve Adventure," "Nutcracker and the King of Mice," "Sand-Man," and others. Vigorous characterizations of highly eccentric personalities, remarkably imaginative situations, and intensely fast pacing has made these tales popular all over the world for 150 years. Editor's introduction. 7 drawings by Hoffmann. xxxiii + 419pp. 5⅜ x 8½.
21793-0 Paperbound $2.25

GHOST AND HORROR STORIES OF AMBROSE BIERCE,
edited by E. F. Bleiler
Morbid, eerie, horrifying tales of possessed poets, shabby aristocrats, revived corpses, and haunted malefactors. Widely acknowledged as the best of their kind between Poe and the moderns, reflecting their author's inner torment and bitter view of life. Includes "Damned Thing," "The Middle Toe of the Right Foot," "The Eyes of the Panther," "Visions of the Night," "Moxon's Master," and over a dozen others. Editor's introduction. xxii + 199pp. 5⅜ x 8½.
20767-6 Paperbound $1.50

THREE GOTHIC NOVELS, *edited by E. F. Bleiler*
Originators of the still popular Gothic novel form, influential in ushering in early 19th-century Romanticism. Horace Walpole's *Castle of Otranto*, William Beckford's *Vathek*, John Polidori's *The Vampyre*, and a *Fragment* by Lord Byron are enjoyable as exciting reading or as documents in the history of English literature. Editor's introduction. xi + 291pp. 5⅜ x 8½.
21232-7 Paperbound $2.00

BEST GHOST STORIES OF LEFANU, *edited by E. F. Bleiler*
Though admired by such critics as V. S. Pritchett, Charles Dickens and Henry James, ghost stories by the Irish novelist Joseph Sheridan LeFanu have never become as widely known as his detective fiction. About half of the 16 stories in this collection have never before been available in America. Collection includes "Carmilla" (perhaps the best vampire story ever written), "The Haunted Baronet," "The Fortunes of Sir Robert Ardagh," and the classic "Green Tea." Editor's introduction. 7 contemporary illustrations. Portrait of LeFanu. xii + 467pp. 5⅜ x 8.
20415-4 Paperbound $2.50

EASY-TO-DO ENTERTAINMENTS AND DIVERSIONS WITH COINS, CARDS, STRING, PAPER AND MATCHES, *R. M. Abraham*

Over 300 tricks, games and puzzles will provide young readers with absorbing fun. Sections on card games; paper-folding; tricks with coins, matches and pieces of string; games for the agile; toy-making from common household objects; mathematical recreations; and 50 miscellaneous pastimes. Anyone in charge of groups of youngsters, including hard-pressed parents, and in need of suggestions on how to keep children sensibly amused and quietly content will find this book indispensable. Clear, simple text, copious number of delightful line drawings and illustrative diagrams. Originally titled "Winter Nights' Entertainments." Introduction by Lord Baden Powell. 329 illustrations. v + 186pp. 5⅜ x 8½.   20921-0 Paperbound $1.00

AN INTRODUCTION TO CHESS MOVES AND TACTICS SIMPLY EXPLAINED, *Leonard Barden*

Beginner's introduction to the royal game. Names, possible moves of the pieces, definitions of essential terms, how games are won, etc. explained in 30-odd pages. With this background you'll be able to sit right down and play. Balance of book teaches strategy — openings, middle game, typical endgame play, and suggestions for improving your game. A sample game is fully analyzed. True middle-level introduction, teaching you all the essentials without oversimplifying or losing you in a maze of detail. 58 figures. 102pp. 5⅜ x 8½.   21210-6 Paperbound $1.25

LASKER'S MANUAL OF CHESS, *Dr. Emanuel Lasker*

Probably the greatest chess player of modern times, Dr. Emanuel Lasker held the world championship 28 years, independent of passing schools or fashions. This unmatched study of the game, chiefly for intermediate to skilled players, analyzes basic methods, combinations, position play, the aesthetics of chess, dozens of different openings, etc., with constant reference to great modern games. Contains a brilliant exposition of Steinitz's important theories. Introduction by Fred Reinfeld. Tables of Lasker's tournament record. 3 indices. 308 diagrams. 1 photograph. xxx + 349pp. 5⅜ x 8.20640-8Paperbound $2.50

COMBINATIONS: THE HEART OF CHESS, *Irving Chernev*

Step-by-step from simple combinations to complex, this book, by a well-known chess writer, shows you the intricacies of pins, counter-pins, knight forks, and smothered mates. Other chapters show alternate lines of play to those taken in actual championship games; boomerang combinations; classic examples of brilliant combination play by Nimzovich, Rubinstein, Tarrasch, Botvinnik, Alekhine and Capablanca. Index. 356 diagrams. ix + 245pp. 5⅜ x 8½.   21744-2 Paperbound $2.00

HOW TO SOLVE CHESS PROBLEMS, *K. S. Howard*

Full of practical suggestions for the fan or the beginner — who knows only the moves of the chessmen. Contains preliminary section and 58 two-move, 46 three-move, and 8 four-move problems composed by 27 outstanding American problem creators in the last 30 years. Explanation of all terms and exhaustive index. "Just what is wanted for the student," Brian Harley. 112 problems, solutions. vi + 171pp. 5⅜ x 8.   20748-X Paperbound $1.50

SOCIAL THOUGHT FROM LORE TO SCIENCE,
H. E. Barnes and H. Becker
An immense survey of sociological thought and ways of viewing, studying, planning, and reforming society from earliest times to the present. Includes thought on society of preliterate peoples, ancient non-Western cultures, and every great movement in Europe, America, and modern Japan. Analyzes hundreds of great thinkers: Plato, Augustine, Bodin, Vico, Montesquieu, Herder, Comte, Marx, etc. Weighs the contributions of utopians, sophists, fascists and communists; economists, jurists, philosophers, ecclesiastics, and every 19th and 20th century school of scientific sociology, anthropology, and social psychology throughout the world. Combines topical, chronological, and regional approaches, treating the evolution of social thought as a process rather than as a series of mere topics. "Impressive accuracy, competence, and discrimination . . . easily the best single survey," *Nation.* Thoroughly revised, with new material up to 1960. 2 indexes. Over 2200 bibliographical notes. Three volume set. Total of 1586pp. 5⅜ x 8.

20901-6, 20902-4, 20903-2 Three volume set, paperbound $9.00

A HISTORY OF HISTORICAL WRITING, *Harry Elmer Barnes*
Virtually the only adequate survey of the whole course of historical writing in a single volume. Surveys developments from the beginnings of historiography in the ancient Near East and the Classical World, up through the Cold War. Covers major historians in detail, shows interrelationship with cultural background, makes clear individual contributions, evaluates and estimates importance; also enormously rich upon minor authors and thinkers who are usually passed over. Packed with scholarship and learning, clear, easily written. Indispensable to every student of history. Revised and enlarged up to 1961. Index and bibliography. xv + 442pp. 5⅜ x 8½.

20104-X Paperbound $2.75

JOHANN SEBASTIAN BACH, *Philipp Spitta*
The complete and unabridged text of the definitive study of Bach. Written some 70 years ago, it is still unsurpassed for its coverage of nearly all aspects of Bach's life and work. There could hardly be a finer non-technical introduction to Bach's music than the detailed, lucid analyses which Spitta provides for hundreds of individual pieces. 26 solid pages are devoted to the B minor mass, for example, and 30 pages to the glorious St. Matthew Passion. This monumental set also includes a major analysis of the music of the 18th century: Buxtehude, Pachelbel, etc. "Unchallenged as the last word on one of the supreme geniuses of music," John Barkham, *Saturday Review Syndicate.* Total of 1819pp. Heavy cloth binding. 5⅜ x 8.

22278-0, 22279-9 Two volume set, clothbound $15.00

BEETHOVEN AND HIS NINE SYMPHONIES, *George Grove*
In this modern middle-level classic of musicology Grove not only analyzes all nine of Beethoven's symphonies very thoroughly in terms of their musical structure, but also discusses the circumstances under which they were written, Beethoven's stylistic development, and much other background material. This is an extremely rich book, yet very easily followed; it is highly recommended to anyone seriously interested in music. Over 250 musical passages. Index. viii + 407pp. 5⅜ x 8.

20334-4 Paperbound $2.25

THREE SCIENCE FICTION NOVELS,
John Taine
Acknowledged by many as the best SF writer of the 1920's, Taine (under the name Eric Temple Bell) was also a Professor of Mathematics of considerable renown. Reprinted here are *The Time Stream*, generally considered Taine's best, *The Greatest Game*, a biological-fiction novel, and *The Purple Sapphire*, involving a supercivilization of the past. Taine's stories tie fantastic narratives to frameworks of original and logical scientific concepts. Speculation is often profound on such questions as the nature of time, concept of entropy, cyclical universes, etc. 4 contemporary illustrations. v + 532pp. 5⅜ x 8⅜.

21180-0 Paperbound $2.50

SEVEN SCIENCE FICTION NOVELS,
H. G. Wells
Full unabridged texts of 7 science-fiction novels of the master. Ranging from biology, physics, chemistry, astronomy, to sociology and other studies, Mr. Wells extrapolates whole worlds of strange and intriguing character. "One will have to go far to match this for entertainment, excitement, and sheer pleasure . . ."*New York Times*. Contents: The Time Machine, The Island of Dr. Moreau, The First Men in the Moon, The Invisible Man, The War of the Worlds, The Food of the Gods, In The Days of the Comet. 1015pp. 5⅜ x 8.

20264-X Clothbound $5.00

28 SCIENCE FICTION STORIES OF H. G. WELLS.
Two full, unabridged novels, *Men Like Gods* and *Star Begotten*, plus 26 short stories by the master science-fiction writer of all time! Stories of space, time, invention, exploration, futuristic adventure. Partial contents: *The Country of the Blind, In the Abyss, The Crystal Egg, The Man Who Could Work Miracles, A Story of Days to Come, The Empire of the Ants, The Magic Shop, The Valley of the Spiders, A Story of the Stone Age, Under the Knife, Sea Raiders*, etc. An indispensable collection for the library of anyone interested in science fiction adventure. 928pp. 5⅜ x 8.

20265-8 Clothbound $5.00

THREE MARTIAN NOVELS,
Edgar Rice Burroughs
Complete, unabridged reprinting, in one volume, of Thuvia, Maid of Mars; Chessmen of Mars; The Master Mind of Mars. Hours of science-fiction adventure by a modern master storyteller. Reset in large clear type for easy reading. 16 illustrations by J. Allen St. John. vi + 490pp. 5⅜ x 8½.

20039-6 Paperbound $2.50

AN INTELLECTUAL AND CULTURAL HISTORY OF THE WESTERN WORLD,
Harry Elmer Barnes
Monumental 3-volume survey of intellectual development of Europe from primitive cultures to the present day. Every significant product of human intellect traced through history: art, literature, mathematics, physical sciences, medicine, music, technology, social sciences, religions, jurisprudence, education, etc. Presentation is lucid and specific, analyzing in detail specific discoveries, theories, literary works, and so on. Revised (1965) by recognized scholars in specialized fields under the direction of Prof. Barnes. Revised bibliography. Indexes. 24 illustrations. Total of xxix + 1318pp.

21275-0, 21276-9, 21277-7 Three volume set, paperbound $8.25

HEAR ME TALKIN' TO YA, *edited by Nat Shapiro and Nat Hentoff*
In their own words, Louis Armstrong, King Oliver, Fletcher Henderson, Bunk Johnson, Bix Beiderbecke, Billy Holiday, Fats Waller, Jelly Roll Morton, Duke Ellington, and many others comment on the origins of jazz in New Orleans and its growth in Chicago's South Side, Kansas City's jam sessions, Depression Harlem, and the modernism of the West Coast schools. Taken from taped conversations, letters, magazine articles, other first-hand sources. Editors' introduction. xvi + 429pp. 5⅜ x 8½.   21726-4 Paperbound $2.00

THE JOURNAL OF HENRY D. THOREAU
A 25-year record by the great American observer and critic, as complete a record of a great man's inner life as is anywhere available. Thoreau's Journals served him as raw material for his formal pieces, as a place where he could develop his ideas, as an outlet for his interests in wild life and plants, in writing as an art, in classics of literature, Walt Whitman and other contemporaries, in politics, slavery, individual's relation to the State, etc. The Journals present a portrait of a remarkable man, and are an observant social history. Unabridged republication of 1906 edition, Bradford Torrey and Francis H. Allen, editors. Illustrations. Total of 1888pp. 8⅜ x 12¼.
                    20312-3, 20313-1 Two volume set, clothbound $30.00

A SHAKESPEARIAN GRAMMAR, *E. A. Abbott*
Basic reference to Shakespeare and his contemporaries, explaining through thousands of quotations from Shakespeare, Jonson, Beaumont and Fletcher, North's *Plutarch* and other sources the grammatical usage differing from the modern. First published in 1870 and written by a scholar who spent much of his life isolating principles of Elizabethan language, the book is unlikely ever to be superseded. Indexes. xxiv + 511pp. 5⅜ x 8½. 21582-2 Paperbound $3.00

FOLK-LORE OF SHAKESPEARE, *T. F. Thistelton Dyer*
Classic study, drawing from Shakespeare a large body of references to supernatural beliefs, terminology of falconry and hunting, games and sports, good luck charms, marriage customs, folk medicines, superstitions about plants, animals, birds, argot of the underworld, sexual slang of London, proverbs, drinking customs, weather lore, and much else. From full compilation comes a mirror of the 17th-century popular mind. Index. ix + 526pp. 5⅜ x 8½.
                    21614-4 Paperbound $2.75

THE NEW VARIORUM SHAKESPEARE, *edited by H. H. Furness*
By far the richest editions of the plays ever produced in any country or language. Each volume contains complete text (usually First Folio) of the play, all variants in Quarto and other Folio texts, editorial changes by every major editor to Furness's own time (1900), footnotes to obscure references or language, extensive quotes from literature of Shakespearian criticism, essays on plot sources (often reprinting sources in full), and much more.

HAMLET, *edited by H. H. Furness*
Total of xxvi + 905pp. 5⅜ x 8½.
                    21004-9, 21005-7 Two volume set, paperbound $5.25
TWELFTH NIGHT, *edited by H. H. Furness*
Index. xxii + 434pp. 5⅜ x 8½.
                    21189-4 Paperbound $2.75

LA BOHEME BY GIACOMO PUCCINI,
translated and introduced by Ellen H. Bleiler
Complete handbook for the operagoer, with everything needed for full enjoyment except the musical score itself. Complete Italian libretto, with new, modern English line-by-line translation—the only libretto printing all repeats; biography of Puccini; the librettists; background to the opera, Murger's La Boheme, etc.; circumstances of composition and performances; plot summary; and pictorial section of 73 illustrations showing Puccini, famous singers and performances, etc. Large clear type for easy reading. 124pp. 5⅜ x 8½.
20404-9 Paperbound $1.25

ANTONIO STRADIVARI: HIS LIFE AND WORK (1644-1737),
W. Henry Hill, Arthur F. Hill, and Alfred E. Hill
Still the only book that really delves into life and art of the incomparable Italian craftsman, maker of the finest musical instruments in the world today. The authors, expert violin-makers themselves, discuss Stradivari's ancestry, his construction and finishing techniques, distinguished characteristics of many of his instruments and their locations. Included, too, is story of introduction of his instruments into France, England, first revelation of their supreme merit, and information on his labels, number of instruments made, prices, mystery of ingredients of his varnish, tone of pre-1684 Stradivari violin and changes between 1684 and 1690. An extremely interesting, informative account for all music lovers, from craftsman to concert-goer. Republication of original (1902) edition. New introduction by Sydney Beck, Head of Rare Book and Manuscript Collections, Music Division, New York Public Library. Analytical index by Rembert Wurlitzer. Appendixes. 68 illustrations. 30 full-page plates. 4 in color. xxvi + 315pp. 5⅜ x 8½.   20425-1 Paperbound $2.25

MUSICAL AUTOGRAPHS FROM MONTEVERDI TO HINDEMITH,
Emanuel Winternitz
For beauty, for intrinsic interest, for perspective on the composer's personality, for subtleties of phrasing, shading, emphasis indicated in the autograph but suppressed in the printed score, the mss. of musical composition are fascinating documents which repay close study in many different ways. This 2-volume work reprints facsimiles of mss. by virtually every major composer, and many minor figures—196 examples in all. A full text points out what can be learned from mss., analyzes each sample. Index. Bibliography. 18 figures. 196 plates. Total of 170pp. of text. 7⅞ x 10¾.
21312-9, 21313-7 Two volume set, paperbound $5.00

J. S. BACH,
Albert Schweitzer
One of the few great full-length studies of Bach's life and work, and the study upon which Schweitzer's renown as a musicologist rests. On first appearance (1911), revolutionized Bach performance. The only writer on Bach to be musicologist, performing musician, and student of history, theology and philosophy, Schweitzer contributes particularly full sections on history of German Protestant church music, theories on motivic pictorial representations in vocal music, and practical suggestions for performance. Translated by Ernest Newman. Indexes. 5 illustrations. 650 musical examples. Total of xix + 928pp. 5⅜ x 8½.   21631-4, 21632-2 Two volume set, paperbound $4.50

THE METHODS OF ETHICS, *Henry Sidgwick*
Propounding no organized system of its own, study subjects every major methodological approach to ethics to rigorous, objective analysis. Study discusses and relates ethical thought of Plato, Aristotle, Bentham, Clarke, Butler, Hobbes, Hume, Mill, Spencer, Kant, and dozens of others. Sidgwick retains conclusions from each system which follow from ethical premises, rejecting the faulty. Considered by many in the field to be among the most important treatises on ethical philosophy. Appendix. Index. xlvii + 528pp. 5⅜ x 8½.
21608-X Paperbound $2.50

TEUTONIC MYTHOLOGY, *Jakob Grimm*
A milestone in Western culture; the work which established on a modern basis the study of history of religions and comparative religions. 4-volume work assembles and interprets everything available on religious and folkloristic beliefs of Germanic people (including Scandinavians, Anglo-Saxons, etc.). Assembling material from such sources as Tacitus, surviving Old Norse and Icelandic texts, archeological remains, folktales, surviving superstitions, comparative traditions, linguistic analysis, etc. Grimm explores pagan deities, heroes, folklore of nature, religious practices, and every other area of pagan German belief. To this day, the unrivaled, definitive, exhaustive study. Translated by J. S. Stallybrass from 4th (1883) German edition. Indexes. Total of lxxvii + 1887pp. 5⅜ x 8½.
21602-0, 21603-9, 21604-7, 21605-5 Four volume set, paperbound $11.00

THE I CHING, *translated by James Legge*
Called "The Book of Changes" in English, this is one of the Five Classics edited by Confucius, basic and central to Chinese thought. Explains perhaps the most complex system of divination known, founded on the theory that all things happening at any one time have characteristic features which can be isolated and related. Significant in Oriental studies, in history of religions and philosophy, and also to Jungian psychoanalysis and other areas of modern European thought. Index. Appendixes. 6 plates. xxi + 448pp. 5⅜ x 8½.
21062-6 Paperbound $2.75

HISTORY OF ANCIENT PHILOSOPHY, *W. Windelband*
One of the clearest, most accurate comprehensive surveys of Greek and Roman philosophy. Discusses ancient philosophy in general, intellectual life in Greece in the 7th and 6th centuries B.C., Thales, Anaximander, Anaximenes, Heraclitus, the Eleatics, Empedocles, Anaxagoras, Leucippus, the Pythagoreans, the Sophists, Socrates, Democritus (20 pages), Plato (50 pages), Aristotle (70 pages), the Peripatetics, Stoics, Epicureans, Sceptics, Neo-platonists, Christian Apologists, etc. 2nd German edition translated by H. E. Cushman. xv + 393pp. 5⅜ x 8. 20357-3 Paperbound $2.25

THE PALACE OF PLEASURE, *William Painter*
Elizabethan versions of Italian and French novels from *The Decameron*, Cinthio, Straparola, Queen Margaret of Navarre, and other continental sources — the very work that provided Shakespeare and dozens of his contemporaries with many of their plots and sub-plots and, therefore, justly considered one of the most influential books in all English literature. It is also a book that any reader will still enjoy. Total of cviii + 1,224pp.
21691-8, 21692-6, 21693-4 Three volume set, paperbound $6.75

THE WONDERFUL WIZARD OF OZ, *L. F. Baum*
All the original W. W. Denslow illustrations in full color—as much a part of "The Wizard" as Tenniel's drawings are of "Alice in Wonderland." "The Wizard" is still America's best-loved fairy tale, in which, as the author expresses it, "The wonderment and joy are retained and the heartaches and nightmares left out." Now today's young readers can enjoy every word and wonderful picture of the original book. New introduction by Martin Gardner. A Baum bibliography. 23 full-page color plates. viii + 268pp. 5⅜ x 8.
20691-2 Paperbound $1.95

THE MARVELOUS LAND OF OZ, *L. F. Baum*
This is the equally enchanting sequel to the "Wizard," continuing the adventures of the Scarecrow and the Tin Woodman. The hero this time is a little boy named Tip, and all the delightful Oz magic is still present. This is the Oz book with the Animated Saw-Horse, the Woggle-Bug, and Jack Pumpkinhead. All the original John R. Neill illustrations, 10 in full color. 287pp. 5⅜ x 8.
20692-0 Paperbound $1.75

ALICE'S ADVENTURES UNDER GROUND, *Lewis Carroll*
The original *Alice in Wonderland*, hand-lettered and illustrated by Carroll himself, and originally presented as a Christmas gift to a child-friend. Adults as well as children will enjoy this charming volume, reproduced faithfully in this Dover edition. While the story is essentially the same, there are slight changes, and Carroll's spritely drawings present an intriguing alternative to the famous Tenniel illustrations. One of the most popular books in Dover's catalogue. Introduction by Martin Gardner. 38 illustrations. 128pp. 5⅜ x 8½.
21482-6 Paperbound $1.00

THE NURSERY "ALICE," *Lewis Carroll*
While most of us consider *Alice in Wonderland* a story for children of all ages, Carroll himself felt it was beyond younger children. He therefore provided this simplified version, illustrated with the famous Tenniel drawings enlarged and colored in delicate tints, for children aged "from Nought to Five." Dover's edition of this now rare classic is a faithful copy of the 1889 printing, including 20 illustrations by Tenniel, and front and back covers reproduced in full color. Introduction by Martin Gardner. xxiii + 67pp. 6⅛ x 9¼.
21610-1 Paperbound $1.75

THE STORY OF KING ARTHUR AND HIS KNIGHTS, *Howard Pyle*
A fast-paced, exciting retelling of the best known Arthurian legends for young readers by one of America's best story tellers and illustrators. The sword Excalibur, wooing of Guinevere, Merlin and his downfall, adventures of Sir Pellias and Gawaine, and others. The pen and ink illustrations are vividly imagined and wonderfully drawn. 41 illustrations. xviii + 313pp. 6⅛ x 9¼.
21445-1 Paperbound $2.00

Prices subject to change without notice.

Available at your book dealer or write for free catalogue to Dept. Adsci, Dover Publications, Inc., 180 Varick St., N.Y., N.Y. 10014. Dover publishes more than 150 books each year on science, elementary and advanced mathematics, biology, music, art, literary history, social sciences and other areas.